CONSIDERING CLIMATE CHANGE

In *Considering Climate Change*, Kimberley R. Miner focuses on what will happen in the next 20, 40, and 60 years around the planet and looks at how we can take an active role in planning for a future we had hoped to avoid.

Each chapter is framed around a central concern that will be familiar to all those thinking about climate change and suffering the eco-anxiety that such an enormous challenge can trigger. Miner carefully unpacks these concerns, walking the reader through issues such as future economics and investing, the housing market, food availability, water availability, and infrastructure and pollution impacts. Each chapter also includes input from experts, including a farmer, a glaciologist, and a Wall Street executive, who guide the reader through their best understanding of the future and how to prepare for it.

Considering Climate Change can either be read cover to cover or with a focus on the specific chapters that will help the reader understand the challenges they are currently facing. Either way, the goal is to walk away with a better understanding of how to thrive in this changing world, and not just have hope for the future—but to have a plan.

Kimberley R. Miner is a scientist, professor, author, and motivational speaker who has traveled to the Earth's most extreme environments to understand how the planet is changing. Her team's work has reached audiences globally through BBC News, *The New York Times*, *GQ*, The Today Show, and a Guinness World Record.

Changing Planet

Changing Planet publishes books addressing some of the most critical and controversial issues of our time relating to the environment and sustainable living. The series covers a broad spectrum of topics, from climate change, conservation, and food, to waste, energy, and policy, speaking to the varied and pressing issues that both human and non-human animals face on our planet today.

The Avocado Debate
Honor May Eldridge

Why Biodiversity Matters
Nigel Dudley

Considering Climate Change
Kimberley R. Miner

For more information about this series, please visit: www.routledge.com/Changing-Planet/book-series/CPL

CONSIDERING CLIMATE CHANGE

KIMBERLEY R. MINER

Routledge
Taylor & Francis Group
LONDON AND NEW YORK

Designed cover image: © Getty Images

First published 2026
by Routledge
4 Park Square, Milton Park, Abingdon, Oxon OX14 4RN

and by Routledge
605 Third Avenue, New York, NY 10158

Routledge is an imprint of the Taylor & Francis Group, an informa business

© 2026 Kimberley R. Miner

British Library Cataloguing-in-Publication Data
A catalogue record for this book is available from the British Library

ISBN: 978-1-032-45298-2 (hbk)
ISBN: 978-1-032-45299-9 (pbk)
ISBN: 978-1-003-37632-3 (ebk)

DOI: 10.4324/9781003376323

Typeset in Joanna
by codeMantra

For the Wild Geese.

CONTENTS

1

CONSIDERING THE HISTORY
OF CLIMATE SCIENCE

Recognition of Place: Ventura, California, 1 pm. Eighteen Celsius (65 Fahrenheit) and windy (23 mph to the East). Full sun, no clouds, I can smell the sea.

Imagine the world was ending. Ok, that's way too dramatic. Imagine the planet you live on was changing irrevocably. And you knew. And so, you told everyone (obviously). But no one listened. Not only did they not listen, but they also armed themselves against you—seeking to prove that you were crazy and unreliable. That's what it is like to be a climate scientist in the 20th and 21st centuries.

Three years after I was born, NASA climate scientist James Hansen's speech to the American Congress was broadcast worldwide. He described how climate change could become the world's great existential threat.[1] Thirty-five years later, we have made very little progress in slowing climate warming.

I am now part of this huge fight that has made such small gains. The fight to slow the Earth's warming and thereby slow all of the many consequences. During my lifetime, one thing has become crystal clear to me— that as a species, we are racing against the clock for both climate change and ecosystem collapse—and losing. As a result, everything I have done in my working life—doing fieldwork everywhere from Antarctica to Mt. Everest looking for answers, observing the world, spending my days off filling my jeans pockets with plastic at the seaside—has been in service of listening to and protecting the planet. And little by little, it is making a difference.

DOI: 10.4324/9781003376323-1

To understand what is happening now, we must understand the past. So, let's start from the beginning. Back in 1856, the often-forgotten American scientist Eunice Foote identified that the gas carbon dioxide (CO_2) had warming potential in enclosed spaces—in what she called the 'greenhouse effect.'[2] Though she wasn't allowed to present her work publicly (because she was a woman), and it was later plagiarized (because she wasn't allowed to present her work), it set the background for much of the climate research that would come from the Western world.

In 1896, Swedish scientist Svante Arrhenius applied Eunice Foote's research to the Earth's atmosphere, arguing that changes in atmospheric carbon could alter the entire planet's surface temperature.[3] By this point in human history—during Europe and the United States' industrial revolutions— humans were already notably increasing the atmosphere's carbon content through coal-burning and industrialization.

Fast forward to 1965, when scientists from across scientific disciplines wrote a report for President Johnson in the United States about the growing 'greenhouse effect' and its potential to alter ecosystems irrevocably.[4]

For context, this would be like the council on Superman's home world of Krypton trying to explain what that Krypton was on the edge of disasters years before they sent Kal-El (Superman) and his cousin into the depths of space. This was THAT meeting. President Johnson didn't act, either. Just like the elders from Superman's home world.

In 1977, the oil and gas company Exxon's own scientists reported internally that warming was indeed driven by fossil fuel combustion and would cause planetary-scale changes.[5] They kept this research a secret until it was leaked in 2022 and finally published in the peer-reviewed literature in 2023.[5] These Exxon scientists developed some of the most groundbreaking climate science at the time, never releasing their results. I wonder how much further ahead we would be in our scientific understanding of climate change impacts if that information had been shared with the global scientific community.

Finally, in 1988, James Hansen (I mentioned him already) testified before Congress about his dismay that so little attention was paid to the rapidly growing issue of climate change. The NASA scientist shared that "…the Earth is warmer in 1988 than in the history of measurements…. It (carbon emission) is changing our climate now."[1]

Ultimately, all scientific theories or forecasts about the future under a changing climate will only be partially correct because the future is always shifting. We are all shaping the future right this second. And the next. Once a system (like the Earth's biosphere) transitions to a new steady state (warmer, carbon-rich atmosphere), all parts of the Earth system will find a new steady state. Forests, oceans, and even the wind will behave differently. And shifting back to a previous state will be much harder, or impossible. In 2025, we are seeing the Earth system change as wind patterns start to slow or stagnate over certain regions, and we have droughts and atmospheric rivers that persist longer than they have in the past. This is very scary to live through. What we do know is that with the amount of carbon (as CO_2 and CH_4) humans are dumping into the atmosphere, we are on track for a significant amount of atmospheric warming. With Earth's many complicated systems in play, a warmer planet simultaneously means drier, stormier, wetter, hungrier, and sicker planet. It is a very troubling track to be on.

Only a few years after James Hansen's testimony, as a small girl, I sat in my backyard under a dying aspen tree and let the ladybugs crawl on me while I read stories of other small girls who saved the world from wrinkles across space-time.

The difference between me and the stories I loved in my childhood is that like many of you, I never really wanted to deal with these problems. I wanted my life to be safe, quiet, and peaceful.

I didn't want to wake up and read headlines about which species were on the brink of collapse today. Or fight for science to be recognized in the popular press. Or travel to distant places for field work for months at a time, exhausted and grateful, but sometimes lonely.

I wanted to sit under a tree and read or teach Kindergarten. Or farm and write poetry. Instead, I am working on finding solutions, sleeping rarely, lecturing executives, comforting students, and trying to devise an action plan for how to prepare for the changes to come. Like thousands of others all over the world.

A world-famous climate scientist advised me not to write this book because "your generation doesn't read—Do a TikTok." Here we are, facing a future of worsening climate change, and I am told to talk about it over TikTok in 15–30 second clips.

Despite her comments about the shortfalls of 'my generation' the Millennial and Z generations have a concrete, intuitive understanding of the urgency of climate change. This is why references to climate change (or zombie apocalypses) are on every streaming service, in many movies, and in science fiction novels. I watch them all, too. It comforts us somewhat to see our darkest thoughts processed on the big screen and solved for the benefit of all.

Fairbanks, Alaska, 2022

The goal of this book is to provide a framework for how to think about planning your life during a time of significant climate change. The topics are based on questions I have received many times while speaking to audiences about climate change, and suggest ways to think about the problems, and ideas for how to confront and incorporate these challenges into your life. I have been thinking about these topics for more than a decade—and it has led me to pursue master's degrees and PhDs so that I could understand and better engage with understanding the risks and impacts of climate change to our world. As a climate scientist, I speak with audiences as diverse as elementary school children and investors, working to communicate the most thoughtful and up-to-date science on climate change.

Growing up with the internet, extraordinary novels and art, and unique access to Earth's remote places make the younger generations of the 21st century the most well-informed in human history. Which is lucky, I suppose, because we are being asked to deal with what is arguably the most significant challenge humanity has ever faced.

In every lecture and interview I give, I hear questions and nervous anxiety about how to live in a time of drought, terrible floods, increasing pollution, pandemics, biodiversity loss, the sixth great extinction, and general suffering. With so much changing, it's pretty hard right now. Many wonder: what will the world be like when the Earth is hot and dry when there are catastrophic storms every year? What is the future?

So, what we do now, matters. It is everything.

Our lives will tell some of the greatest stories in the history of humanity. We are already the post-9/11, post-Columbine/Sandy Hook, post-pandemic, multi-war generations, looking down the barrel at the biggest global ecological disaster in human history. Maybe the history books about us will be inspiring tales for future children living in homes that generate their own electricity and water—with vines and morning glories climbing up the walls and shining in the sun. Or maybe the future will be lost in the endless desert of destruction we all fear. I don't know which way it will go.

What I *do* know is that our generation will deal with the impacts of climate change for the rest of our lives. The only choice now is to face the reality of climate change together with courage and strength. The future has *always* been unknowable. At this moment in history, we can choose to care for ourselves and all the plants, animals, and people in our lives. Applying choice and care changes everything. I hope that this book can be a guide through how to put attention and courage into planning for the realities of this uncertain future, a future we had all hoped to avoid.

I know that we must do more than just have 'hope' for the future. We must all find it in ourselves to do the hard things—because we know we are worth it. This planet is worth it. Hearing the birds sing at sunrise each day is worth it. And that is what is on the line.

For the rest of the book, I'll share my conversations with some of the innovative minds of their fields to help us understand how to deal with this weird and scary future under climate change. From economics and investing to housing and family planning, the experts I've spoken to will talk us through their understanding of the future. There are ideas for how you can be part of the solution throughout. You can read this book cover to cover or just focus on the chapters that interest you. They are a series of short essays based on the current state of thought. I hope that these small conversations will help you walk away with a better understanding of how to **thrive** in this changing world. I hope the ideas provide some comfort in the turmoil that always arises during change. I am writing from my perspective, so the focus is derived from my personal experiences, and those of the people I have encountered on my travels.

At this time on the planet, we must decide how we want to be in relationship with each other, how to successfully live the best dreams of

Svalbard, 2024

sustainability and sufficiency, to share the dreams of everyone who came before us, to be honorable and considerate of those on the Earth with us, and to preserve the wonderfulness of the world for those who come after us. It is time to take action and have a plan for the future. We are the dreamers, the believers, and the creators that will decide, and make our future on Earth. We have no other choice. So, let's begin this journey together.

REFERENCES

1 Shabecoff, P. Global Warming Has Begun, Expert Tells Senate. *The New York Times* (1988). https://www.nytimes.com/1988/06/24/us/global-warming-has-begun-expert-tells-senate.html.

2 The Woman Who Demonstrated the Greenhouse Effect. *Scientific American*. https://www.scientificamerican.com/article/the-woman-who-demonstrated-the-greenhouse-effect/.

3 The Nobel Prize in Chemistry 1903. *NobelPrize.org*. https://www.nobelprize.org/prizes/chemistry/1903/arrhenius/biographical/.

4 Statement by the President in Response to Science Advisory Committee Report on Pollution of Air, Soil, and Waters. *The American Presidency Project*. https://www.presidency.ucsb.edu/documents/statement-the-president-response-science-advisory-committee-report-pollution-air-soil-and.

5 Supran, G., Rahmstorf, S. & Oreskes, N. Assessing ExxonMobil's Global Warming Projections. *Science* **379**, eabk0063 (2023).

2

CONSIDERING MENTAL HEALTH AND CLIMATE CHANGE

Los Angeles, California, 9 pm. Nineteen degrees Celsius (66 Fahrenheit), no wind, big dipper setting. I smell gyoza and cigarette smoke.

It's alright to feel sad or scared about climate change. You can't have grief without love. Just as we all love the planet we all live on.

Many climate change books that tell you to feel optimistic ignore the reality that I saw three birds dead from the heat outside my apartment in the Los Angeles summer yesterday. And on any given day, many I work with in climate science are heartbroken, livid, and melancholy in alternating phases. So, it's ok to feel sad. To feel it for all the animal people, plant people, and human people on the planet right now.

Little by little, try to remember that none of us are alone. We are all part of a living ecosystem that can heal and improve itself. We can deepen our relationship with the world around us, whether taking the seeds that are becoming tree seedlings from the small yard to plant by the river or feeding the birds who are migrating over a city of millions, who fly high while looking for food and water. Building strong relationships with the world around me is the optimism I believe in. The optimism born of wholehearted action. Of small, kind gestures. And that is where we will start.

Around the world, kind-hearted people are working diligently, against all odds, trying to predict the future. When my mentor at NASA JPL won the Nobel Peace Prize with Al Gore and the other IPCC scientists in 2007,[1] there was a strong sentiment that if the threat of climate change could be

DOI: 10.4324/9781003376323-2

communicated properly, the world would be compelled to act. But slowly, they learned that confronting someone with knowledge and data about our changing planet may challenge their self-identity and their feeling of place in the world. Bringing someone the evidence that makes them feel responsible for a problem as big as the whole world feels like self-annihilation. It is very unsettling. Now, 15 years and very little action later, my mentor is considering the very plausible future we will experience under the most extreme warming scenario.[2] Today, though, the ocean and forests are still absorbing much of the carbon humans produce. We are overly relying on them to manage the carbon we produce, and these delicate systems are beginning to feel the strain.[3]

Now, my mentor and scientists like him are worrying about what happens when these very large natural systems stop absorbing carbon. When the forests, Arctic permafrost, or oceans shift and start to release the CO2, methane (CH4), or other greenhouse gases than they absorb, they will be joining humans in pushing carbon into the atmosphere and tipping the scales toward instability.[3] Once that happens, the question becomes: can our Earth system ever return to the world we have known?

If you speak with one of these scientists who have been alerting the world to climate change impacts for decades, it is unsettling to hear the struggles they have had to fight. When my mentor worked at a notable science research center decades ago, he was told there was no reason for a research program on carbon—it was an element that wasn't notable in any specific way. Years later, when they posted his Nobel recognition on the wall, he had a rare moment of satisfaction. I say rare because many climate scientists spend their entire lives trying to explain that climate change could be one of the most threatening, notable, and entirely preventable challenges in human history.

Today, many thousands more people are working diligently to understand the fallout of climate change across the world. I've heard from many people who are changing industries and careers and moving houses or cities to be a part of working to solve the climate catastrophe.

My friend from a First Nation in the Yukon calls this period the 'most recent climate catastrophe' as a reminder that the people who have lived with the land for millennia have seen many changes. What is different right now is how rapidly the change is happening. This friend moved away from a career in mining to use his chemistry background to manage his ancestral homeland. His team is adapting generations of traditional knowledge to forecast and manage future climate changes throughout their territory.

Another friend works in environmental remediation. She goes to the most polluted sites in the nation (superfund sites) to monitor progress to clean the air, water, and soil. She travels with men twice her age who leer at her. She stays in shabby, dirty hotels, and eats poor quality, processed food as she moves city to city, remediating PFAS, oil spills, nuclear waste dumps, and all those other chemicals that humans have released without cause or a plan for cleaning up. She is lonely. And she is driven. There are hundreds of people doing this work. And they are all exhausted. But no one, no one, is giving up.

Recent research shows that the cause of generalized climate-anxiety isn't seeing the evidence from the scientific community, the reports from the IPCC, or any other type of science reports.[4] It is the lived *experience* of climate change we are all having.

As humans, we know in our bones that something is wrong with the ecosystems around us—that our relationship with our planet is not well. The years 2022, then 2023, and now 2024 were the hottest years on record around the globe. Temperatures not predicted until 2050 have already hit Europe and Asia multiple times. The Rhine and the Colorado rivers dried up in places, and floods from atmospheric rivers hit Pakistan and China. In 2023, unseasonal rainfall flooded California. The Swiss Alps got little winter snow, and a record drought devastated many African countries. Pictures of dead giraffes and struggling farmers were strung across the internet. New oil drilling sites in the Arctic led to a public outcry, and there is ongoing debate about which ecosystem the next pandemic virus will come from. The looming question behind every natural disaster headline seems to be— which destroyed ecosystem will come for humans next? These disasters are not natural anymore, they are caused by fossil fuel use, and are happening far sooner than Western scientists predicted.

On top of all that, after experiencing a climate (or fossil fuel) disaster, up to 40% of people report experiencing post-traumatic stress disorder (PTSD), where their nervous system is thrown into chaos.[5] With PTSD, it is difficult to understand one's place in the world. The ground feels unstable beneath your feet. With the number and severity of these disasters increasing every year, the diverse and serious consequences for the human population will continue to grow.

The Earth system we are a part of is incredibly complicated, and it is difficult to predict or understand the scale of what is going wrong. As the fallout from generations of pollution and waste continues to hit us, it's ok to feel sad, furious, and worried, or any other emotion. I feel it too.

Parents reach out to me weekly, sometimes daily, to tell me that their children are worried to tears. I see people younger than me marching down the streets during school hours and hear the concern when I speak with student groups all over the world on Zoom. These worries and the hope that I can provide some small comfort keep me talking to as many student groups as will have me.

Dr. Renee Lertzman[6] is a climate psychologist focusing on the grief of the climate catastrophe. When I spoke with her, she was in the San Francisco Bay area, the ancestral homeland of Ramaytush Ohlone. It was Tuesday, just under 70 degrees Fahrenheit (21 degrees Celsius), and perfectly sunny.

I ask her how to deal with the pain and struggle of eco-anxiety.

Dr. Lertzman says that naming your feelings is the first step in dealing with something as overwhelming as feelings about climate change. **Naming the feeling you are feeling.** That could be eco-anxiety, climate grief, rage, anger, anxiety, or just sadness. Just call it something, so that you can hold the feeling out and look at it.

The next step is **accepting these feelings and realizing they are normal.** They are a logical response to what is happening around us. It is normal to feel strongly about something as all-encompassing as climate change. It is normal to be worried when the planet around you is changing and many species are dying. Feeling eco-anxiety means that you are paying attention; it is your body protecting you and showing your desire to protect the Earth. Something is wrong on the planet, and you have noticed it and feel compelled by it. In that way, these feelings are a gift.

Part of what is so hard about climate-anxiety is that this is a relatively new emotional state for humanity. We have never been so acutely aware of the state of the entire planet, nor have we ever had so much power to change the world around us. The knowledge that the planet will be essentially and fundamentally different decade by decade within our lifetimes, and that this existential threat is so difficult to fight, is a new and terrifying emotional, physical and mental state for us. It's important to take time to explore those feelings and create spaces to **explore your feelings.** Talking about your feelings in community helps to process the potentially overwhelming thoughts we may have about our changing planet. There are online and in-person resources popping up all over the world to do this— whether in coffee shops and tea houses or schools, there are increasingly more communities focused on holding space for feelings about climate change.

I believe that our climate-anxiety is a sign that we are so closely attuned to the planet around us, that we can feel the changes happening.[7] And that connection, that closeness we feel with this giant biosphere traveling through the vast space of galaxy, is a gift. It is a blessing to be this tied to the Earth we live on. And I hope the pain and fear we are experiencing will propel us to action and creativity in service to a beautiful future. Taking action is key to caring for our mental health.

The next step is to make a choice to **do something**. Do something to further integrate yourself with the world around you. Not by having optimism or ignoring the signs in our bodies and ourselves, but by strengthening our relationship with the natural world to honor and prioritize these feelings. This can be acknowledging this beautiful day, providing food for your wild neighborhood birds, going for a walk and picking up litter, or just sitting outside feeling your feet on the ground and the air in your lungs. Studies have shown that being outside can greatly improve health in a variety of ways, including mental, physical, and even heart health.[8]

Our modern society convinces us to elevate our relationships with money and stuff. To heal our climate-anxiety, psychologists like Dr. Lertzman say we must put our relationship with all living things back at the center of our experience. Bringing our relationship with the living world into focus can be extraordinary. It can be like waking up to the world around you or like seeing the world as a child again. Every day, even today, you can take small steps to remember that you are forever part of the Earth. If you have the space and finances, plant some native or pollinator plants. Write a letter about your concerns to your political representatives. Pick up garbage at a local park, beach, neighborhood, or even a parking lot. Put out water for the animals and turn off your outdoor lights at night. We can take hundreds of small steps to ameliorate our climate anxiety while keeping our sights set on the long-term goal of a complete transition away from ongoing carbon emissions (for more on small steps, check out the UN[9]).

These actions at an individual level will not change the state of the world on their own. Let's not confuse these small actions of integration and kinship with fixing the climate crisis. As many of my scholar friends focusing on decolonialism note—the system is working exactly how it is supposed to. In colonial societies, the big companies take credit for the positive parts of life and blame communities for the negative. This system of thinking makes you feel powerless and alone. It makes you take responsibility for things that are not your fault. In true surrealist fashion, you are not supposed to understand what the problem really is—and even what is not. That is the point. Spreading false news and despair at our choices is part of how fossil fuel companies have done so well financially, even as they researched and mapped out the warming of the world.[10]

But all of us, together, cannot be pushed aside. We can make this moment in human history a watershed, we can become guides for the next generation. We can make this a time of collective action where the small streams become great rivers. A time when we can choose to relate to the world around us differently. We need to create stories that invest in a new, or old, direction. A world you can feel good about. We can move from transactional relationships to spaces of listening and nurturing, where we prioritize consent and collaboration, between ourselves and across species. Today, when you are done with this chapter, you can take a few small actions to help yourself feel more a part of the world around you.

We must forgive those who came before us for the mistakes they made, and are making, and commit to doing better ourselves. Follow in the footsteps of their good teaching and novel science. We must have the conviction to believe that humans can live peacefully and symbiotically, with all other species, and that we can be a part of making that future. Let any grief, any ecoanxiety you have drive small daily actions, a life *dedicated* to improving the future. A life of change through thousands of small steps, taken together.

Steps to healing climate grief, in summary:

- Name what you are feeling
- Feel it
- Accept that this is a normal, common feeling that shows a deep connection to the planet
- Do something (take action)

A FEW RESOURCES FOR FURTHER SUPPORT

Renee Lertzman, Accessed 2025. https://projectinsideout.net/the-quadrant-of-engagement/.

Sacha Wright and Emily Osterloff, Accessed 2025. https://www.nhm.ac.uk/discover/how-to-cope-with-eco-anxiety.html.

REFERENCES

1 The Nobel Peace Prize 2007. NobelPrize.org. https://www.nobelprize.org/prizes/peace/2007/summary/.

2 Schimel, D. S. & Carroll, D. Carbon Cycle–Climate Feedbacks in the Post-Paris World. *Ann Rev Earth Planet Sci* **52**, 467–493 (2024).

3 Lenton, T. M., Rockström, J., Gaffney, O., Rahmstorf, S., Richardson, K., Steffen, W. & Schellnhuber, H. J. Climate Tipping Points – Too Risky to Bet Against. *Nature* **575**, 592–595 (2019). https://www.nature.com/articles/d41586-019-03595-0.

4 Gergis, J., Blashki, G., Gardner, J. & Bradshaw, S. *Climate Trauma: The Growing Toll of Climate Change on the Mental Health of Australians* (The Climate Council of Australia Limited, 2023). ISBN: 978-0-6450500-8-0 (print) 978-0-6450500-9-7 (ebook). https://www.climatecouncil.org.au/wp-content/uploads/2023/02/Report-Climate-Change-and-Mental-Health.pdf?utm_source=substack&utm_medium=email.

5 Goldmann, E. & Galea, S. Mental Health Consequences of Disasters. *Annu Rev Public Health* **35**, 169–183 (2014).

6 Lertzman, R. Renée Lertzman | Speaker | TED. https://www.ted.com/speakers/renee_lertzman.

7 Ashlee, C., Harperet, S. L., Minor, K., Hayes, K., Williams, K. G. & Howard, C. Ecological Grief and Anxiety: The Start of a Healthy Response to Climate Change? *Lancet Planet Health* **4**, e261–e263 (2020). https://www.thelancet.com/journals/lanplh/article/PIIS2542-5196%2820%2930144-3/fulltext.

8 3 Ways Getting Outside Into Nature Helps Improve Your Health. *Cultivating-Health*. https://health.ucdavis.edu/blog/cultivating-health/3-ways-getting-outside-into-nature-helps-improve-your-health/2023/05.

9 United Nations. Accessed 2024. https://www.un.org/en/actnow/ten-actions.

10 Supran, G., Rahmstorf, S. & Oreskes, N. Assessing ExxonMobil's Global Warming Projections. *Science* **379**, eabk0063 (2023).

3

CONSIDERING FOOD AND CLIMATE CHANGE

5:05 pm local time, New York City, at the corner of 34th and 8th. Tesla lived in this building until his death. Clear, sunny, fall breeze and 21 degrees Celsius (70 Fahrenheit).

We are in the middle of the sixth great extinction. For the first time in history, animals and plants are going extinct in record numbers because of human activity. The World Wildlife Fund notes that over 40% of land on Earth has been converted to farmland for humans,[1] a trend that decreases wildlife habitat and exposes ecosystems to toxic chemicals. Let's talk about the hows and whys.

My bachelor's degree was in a part of ecology called agroecology[14]— broadly, the application of ecosystem thinking (understanding the relationships between plants, animals, the physical environment) to farming. When it is successful, agroecological food systems can even have a positive impact on the environment around them. This may be made possible by increasing the availability of clean water, shade, and habitat for pollinators and animals. They may also be able to produce higher yields than industrial farms, meaning that the advancements in farming in the last hundred years have resulted in a high-producing, ecosystem-forward way of thinking about farming.

The modern, industrialized, food system began to evolve into its current state just during and just after World War 2.[2] There were many mouths to feed, many destabilized countries, and a plethora of new technologies that had been developed for the war.[3] Some of these technologies included chemicals that had been used in the trenches of the war, to kill insects, and

DOI: 10.4324/9781003376323-3

as weapons. Since the technology was now newly available, many of these chemicals were transitioned to killing insects or unwanted plants for an industrial food system responsible for feeding millions, then billions.[4,5] They became commonly used pesticides.

To regain the losses to land and people from the war, this new food system prioritized quantity over quality. It organized rows and rows of one type of food—be it carrots, tomatoes, or corn, instead of mixing all different types of plants and food crops together in one area like in a modern home garden. This idea to put all of one type of plant together on a large-scale farm has since been termed a 'monoculture.' A novel, large-scale, industrial farming regime allowed for technological and productivity breakthroughs unlike the world had ever seen. It could be managed with equipment for picking the crops and preparing the soil—on larger scales than the small farms of generations past.

With subsidies from governments and this new chemical technology in hand, agri-business and industrial farming took hold across the world. Small family farms were no longer required to produce every type of vegetable, animal food, and meat that was needed for each community; now, farms could focus on just producing and shipping one type of food. Their crop (say, carrots) would be brought together with the produce from other large-scale farms in supermarkets. These markets could provide any type of food you wanted, whenever you wanted it. The carrots from one farm in California would sit next to tomatoes from Guatemala and coffee from Northern Africa. An evolving global food trade scene bolstered this global market-type food network, bringing mangos north in February and grains south in the summer.[6]

There were, and are, obviously significant benefits to this type of system. With enough fertilizer and weed killers, an availability and constant supply of food made it possible for the global human population to more than triple since the 1950s. Industrial agriculture contributed greatly to national economies and made any type of food nearly instantly available in many countries.

It is the unforeseen costs and losses of this agricultural system that caused the greatest problems. Other large-scale impacts come from the process of growing food. When it rains on a field covered in pesticides and fertilizer, these chemicals and nutrient additives make their way into the local ecosystem. The chemical pesticides travel through the watershed, through small runoff streams into bigger rivers, and through estuaries killing critical insect species or creating algal blooms at the coast. These can kill hundreds of aquatic species and create aquatic 'dead zones.' The impacts of the runoff of both pesticides and fertilizer have numerous coastal and riverine ecosystems on the verge of collapse.[8-10] Through both the food we eat and the ecosystems we live in, these chemicals can get into our bodies and cause birth defects, illnesses, allergies, and chronic diseases.

There has been an exponential increase of pesticide development since the 1950s leading to their increasing creation and use. There are currently over 17,000 known types of pesticide for use in the food system, both for plant management and to kill small animals and insects. There are even special regulatory loopholes to allow the use of potentially harmful pesticides before the health and safety of these chemicals has been verified. For example, in the United States, the EPA allows for pesticide use on the land before they are safety tested using a rule called 'conditional registration.'[11] When pesticide ingredients are new, or have recently been developed but not tested, the EPA will determine if there could be a 'reasonable' risk to the environment. However, it is truly impossible to know what damage these chemicals may cause to the planet before they enter the environment, which has led to some of these chemicals now known to cause irreversible harm. Methyl Bromide[12] is a known toxic pesticide that falls under one of these special use exemptions in the United States. It has been phased out in most industrialized nations,[13] but the US started providing exemptions for its agricultural use in the early 2000s, and this cancer-causing chemical is applied liberally to strawberries, in nurseries and even on ornamental flowers. The number and availability of toxic pesticides for common use agricultural is staggering. There may also be hidden costs to the industrial food system, including the release of greenhouse gases, overtaxing the water system, health impacts for pesticide and antibiotic exposure, and land degradation from nutrient loss.[18]

When monoculture farming in one location forces the soil to produce food year after year, vital nutrients are depleted, and the soil begins to lose its structure and ability to hold water. Further, the farmers specializing in one vegetable or grain crop are beholden to the international prices for that crop—meaning that in difficult years when the market for their crop is down, they may make little to no money for back-breaking, 80-hour work weeks.

The 21st-century food system may also overproduce the amount of food that humans actually need. The problem is that food isn't even distributed across the planet. Grocery stores throw away multiple trash cans or dumpsters worth of food every day because it is 'expired.' To manage prices by creating demand, and keep their own families fed and housed, farmers sometimes have to burn or leave crops in the field. For example, in 2022, humans produced 38% more food than could have been consumed if it was evenly distributed across the planet. At time of writing, around one-third of food grown for human consumption globally is wasted. About 14% is lost before it is purchased, and an additional 20% wasted in households.[7] Additionally, the process of food loss and waste generates up to 10% of global carbon emissions—about five times the amount generated by planes. The growing problem is that our food production capabilities have grown faster than the amount of food the population needs, leaving us to throw away critical resources while depleting the soils and water crops depend on. It turns out that despite the many gains, the costs of this industrial food system are profound— and still increasing.

As a part of my agroecology education, I lived on and helped manage a small working farm. Ours was a diverse garden with hawks, golden eagles, gophers, and squirrels everywhere. There were never any chemicals applied to the small crops, but there were many candlelight parties and guitar sessions at night. After I graduated, I went to Idaho for a few months to work on a draft-horse farm that both trained draft horses to pull plows for farming — (like the typical pictures from the Netherlands and Amish country) and also produced alfalfa. My goal was to learn how this small family was using huge draft horses to plow their fields and manage their land. And it was hard. Insanely hard. Rise at the crack of dawn and stop a stampede of horses by steadily swinging a rope in a circle while standing in the middle of the road kind of hard. Filthy floors and no time to clean and going to bed without a shower hard. Pray the rains come but not too much and not too fast, or the fields will wash out, and there will be no food or money for groceries kind of hard. It opened my eyes to the real work of farming—the work that is reliant on the good graces of weather and of the seeds doing well.

So, I can't say that the industrial food system is a villain. It has fed billions and made life somewhat easier for many family farmers. The problem is that the side effects of the system are enormous, and this rigid system is structurally unable to be flexible and adapt to the current stress of climate change. There must be a mid-point, a happy medium, between the transnational, industrial, chemical-based, all-or-nothing agricultural system of the last 60 years and small local farms that have no fallback when the growing seasons are bad, or climate change brings drought in summer and flooding winters.

When I asked on social media to interview farmers who are responding to or thinking about climate change, I got hundreds of responses. I heard from small-scale sheep farmers in California, permaculture specialists in Brisbane, Australia, small farmers in Italy, and family farmers who had inherited their land in France. I was amazed. I knew that there were large numbers of small farms globally, but to see these hundreds of responses on one small social media post made me realize this was just a microcosm of the people working across the world to grow food on a regional scale. I wish I could have had the time to speak with all of them. If those responses are just a small sample from one social media platform, I would venture there are millions of people all over the world working to create a modern, or rather, old (I'll explain) sustainable food system.

To get a perspective on this challenge, I spoke with Morgan Kalber-loh, who lives in the traditional homeland of the Nuche or Ute people, the Colorado Front Range. When I spoke with him, the wind was over 70 mph, and he lamented about how other small-scale farmers would be suffering from the stark mix of wind, snow, and hot days from the month prior. I grew up near here in Boulder, close to where Morgan farms in the area where urban life transitions to rural outside of Denver. He tells me that it's not like Chicago or London where there are expansive suburbs and industrial hinterlands that eventually give way to agricultural lands, it's a more of a gradual transition between the two. Morgan is unique in that he doesn't own or even lease the land that he farms. He has found frequently unutilized tracts of land, that are owned by local residents, large enough for an urban farm. His farm is a no-till, worm-based composting style where he grows heirloom tomatoes and peppers, pea shoots, and specialty greens. He sells these organic vegetables to local restaurants.

When we spoke, what struck me about Morgan's model was that it doesn't require expensive infrastructure, huge startup costs, or even own-ing any land. He has transformed empty plots with dead grass or bare soil that lay at the corners of busy intersections into thriving, food-producing ecosystems. He mentioned that as he transforms these plots into thriving gardens, laying compost, building up the soil, and letting the land thrive, he often sees more birds of prey, foxes, field mice, and songbirds. The land that was once bare, losing nutrients to the surrounding environment as the rain, wind, and sun bore down on the dead grass, was now thriving with life. Morgan makes a living revitalizing small plots of land and turning them into specialized farms. He shares that if he owned his land he would expand these practices further, incorporating permaculture to rewild the land, engagement with the community, higher crop diversity and carbon sequestration. He mentions to me that in many cases, land ownership is a barrier to community healing.

Using the skills, tools, and land they have, so many people are exploring the space around what our food system could look like. These are people are looking to find the intersection between the 'old' system of mixed plant farming on a small scale and the new system of rows and rows of carrots. There are universities and degree programs dedicated to exploring these ideas, and they are really making ground (haha). There is an international community of farmers educated by these regenerative agricultural programs, creating small-scale experiments and farming families.

Species grow better together. Whether it is following the South American traditions of planting three sisters or milpa—corn beans and squash—together[15] (because they nurture each other) or using flowers to trap harmful insects in the rows between strawberries—plants, animals, and people do better together. Separating our food system from the rest of the ecosystem by farming in large plots with only one type of plant does more harm than good over time. Intellectually, it separates us from understanding the side effects of the chemicals we apply and methods we employ, and it can reduce community participation and equity in the food system.

Doing better together—this will be true during climate change, too. When ecosystems are thriving, diverse, and filled with plants and animals, they withstand weather shocks better. Whether it is unseasonable heat, intense rainfall, or cicada blooms, crops that are grown together support each other. This is also true because of inherent physical processes including plants that do nitrogen fixation for other plants, diverse insect pollinators who co-evolved with flower species, or underground mycelia networks that allow mushrooms to talk to each other. Diversifying crops also supports the people who depend on them to make money, during years when the rain or heat one year means that just one type of crop can't survive. Diversity is a key strength for thriving ecosystems and will be a benefit as climate change continues to wreak havoc on our seasonal weather patterns.

If you kill one species, you could kill them all. This is true for bugs, plants, and, eventually, people. The chemicals we have developed for 'targeted' species removal don't just kill one thing. These chemicals may kill gnats but they also kill also bumblebees; they may stop mosquitos but then they end up in breast milk—transferring dangerous chemicals between generations of people and animals. If a crop is doused with chemicals, when it rains, the fish in the nearby river are impacted. Basically, if something is designed to kill, it will do just that, and so, of course, it shouldn't

Alaska North Slope, 2023

be applied wholesale to any plant or food crop. There are thousands of examples of this, but one that many know about is Rachel Carson's Silent Spring, a best-selling expose on the application, and horrifying side effects, of DDT.[16]

As the climate changes, the land will too. As mentioned, diversifying food crops and integrating farming into the local ecosystem will make farms more resilient to shocks like unseasonable rainfall or heat. On Earth, every plant, animal, and insect fill a specific niche essential to the whole eco-system thriving. Butterflies pollinate specific flowers, and moths pollinate others at night. Each country, each small region, and each farm has its own complex ecological working relationships and its own diversity of species. The possibility that an increasing number of small farms in urban or rural spaces could increase the opportunities for wildlife to thrive is very real, and in process. There are urban roof gardens, individuals like Morgan, sustain-able coffee farms that grow coffee intermixed with native plants and birds, community gardens in large cities across Europe and rural hillside farms that are leading the way toward a diverse, mixed system. Creating ecosys-tems to produce food where the wild birds, insects, and the cultivated crops are mutually supportive and will allow all to thrive.

Svalbard, 2024

Food choices make a difference. Many people have been discussing how our food choices can support or harm the changing planet. There seem to be endless permutations of diet choices that have some claim to environmentalism, including going vegan (no animal products), vegetarian (no meat), pescatarian (no meat except fish), eating meat once a week, gluten-free (no gluten grain products), dairy free, organic only, local only, etc...

Maine, 2018

There is, happily, science to support some of the claims that these diet choices make a difference. Eating local, within some number of miles from where you live (often less than 100 miles or ~480km), limits the amount of carbon that is required to get food to you (www.eatforum. org). However, the agricultural practices are also important to consider. For example, does this farm or business spray pesticides? How do they treat their animals?

So, keep it local… The seemingly endless minutiae of decisions about how and where we consume food can feel overwhelming at best and completely not worth considering, at worst. So, to make things simple, the predominant science says that to make a difference in the climate system and to live most sustainably, we should be eating very little meat and industrial dairy, mostly local, and mostly crops that are growing in that season. What we each choose to do with this information is up to us—but let me add one more thing. Choosing local food also protects us as individuals from climate change and other crises. Whether it is an economic downturn, or a sudden storm, disrupted shipping routes and transportation networks for markets can make it difficult to get food in. A blend of regional and local food systems will help protect us from climate shocks.

…and based on your ecosystem. My advisor from my undergraduate degree[17] just had the 25th anniversary of his vineyard in the dry lands outside of Santa Barbara, California. He has brought over an old-world wine grape growing methodology from Italy for his crops: dry farming. What this basically means is that he doesn't water his grapes except in very small quantities at certain times of the year—this allows the grape plant roots find the groundwater on their own. While most vineyards water their grapes every day, using hundreds of thousands of gallons a year, he waters a bit in the spring as needed and then reverts to the old methodology of praying for rain and training the plants. At my mentor's vineyard, the entire community, including us students, helped with the pruning and, eventually, the grape picking to make his wine. This method is a part of the farming system he helped popularize, which is part of agroecology. The dry farming and no-till methods are being applied all over the world, to most efficiently use the soil minerals and clean water we have available. The system basically allows him to grow wine or other food alongside a natural system—without harming the ecosystem. This method looks at the natural structure of the environment and seeks to mimic it. My adviser always reminded us that food ties us forever to the world around us, and the people who grow it. It makes us a part of, and reliant on, our ecosystem.

The re-emergence of agroecology and the integration of more traditional old-world farming styles being taught in schools[3] internationally is also an indicator of a global shift back to incorporating traditional knowledge for growing food in harmony with the ecosystem. This infusion of older farming methods focused and sustainable and regenerative land cultivation reintroduces culturally relevant foods to displaced or colonized communities, it can make food more available, and food systems more equitable as individuals can participate more directly in their food systems. And, as Morgan noted, this localized, traditional-modern fusion that agroecology and similar farming methods provide is possible right now. Right here. Wherever you are.

Small-scale farmers across the planet are working to move more closely with the land, with the seasons, and to provide food at lower cost, higher quality, and with less ecosystem impact. This shift toward older methodologies may mean that our tastes need to change as well. With growing climate impacts, it may not be reasonable for folks in Maine to demand fresh mangos in December. There is a considerable carbon and ecosystem cost to eating out of season—and that may be something we will have to learn to be more careful about in the future. Diversifying our food preferences and moving toward the food choices that are more sustainable, and appropriate for our region could also provide economic and environmental stability—buying local crops from providers who grow diverse, nutritional foods.

A majority of the processed food we see in grocery stores comes from only ten trans-national companies that create, process, and distribute everything from tea bags to breakfast cereal and microwave dinners.[19]

This small number of companies controlling so much food supply is of growing concern during a time of unpredictable climate shocks, and social inequity.

Expanding our tastes to include foods that our ancestors ate, foods local to your ecosystem, and food with a lower carbon footprint (maybe even bugs!) will be a crucial part of lowering both your individual risk from climate change shocks and improving the lives of all of the creatures caught up in our industrial food system. Already, Michelin-star chefs across the globe are thinking about how to bring back the tastes, colors, and smells of their grandparents' generation—modernizing the old ingredients and introducing them again. They remind us that all food used to be 'organic.'

As a world community, we have very tough choices ahead of us about how we get our food, where it comes from, and how it is treated during every step of that process. If we are to start seeing our world as the interconnected beauty it truly is, reassessing the impacts of the industrial food system must be a part of that conversation. Our current methodology for feeding ourselves has destroyed many species, increased the carbon in the atmosphere and led to global warming, polluted environments, and left us vulnerable to ecosystem collapse. There is a better way—and the wonderful part is—we already know what it is.

Things to consider about our changing food system, in summary:

- Animals and plants grow better together
- Even if targeted, killing one species could kill them all
- Climate change is driving many changes to the land and agriculture
- Food choices make a difference
- Keep it local and based on your ecosystem

A FEW RESOURCES FOR FURTHER SUPPORT

Agroecology Knowledge Hub. https://www.fao.org/agroecology/home/en/.

Travel Global, Eat Local. https://www.tasteatlas.com/.

Mapping U.S. Food System Localization Potential: The Impact of Diet on Foodsheds. https://pubs.acs.org/doi/10.1021/acs.est.9b07582.

Gardens: Backyard Farming – A Guide to Homesteading for Beginners. https://www.homesandgardens.com/advice/backyard-farming-homesteading.

REFERENCES

1 What Is the Sixth Mass Extinction and What Can We Do About It? | Stories | WWF. *World Wildlife Fund.* https://www.worldwildlife.org/stories/what-is-the-sixth-mass-extinction-and-what-can-we-do-about-it.

2 Hueston, W. & McLeod, A. Overview of the Global Food System: Changes over Time/Space and Lessons for Future Food Safety. In *Improving Food Safety Through a One Health Approach: Workshop Summary* (National Academies Press, Washington, DC, 2012).

3 State of Oregon: World War II – Farm Labor Programs Work to Bring in the Crops. https://sos.oregon.gov/archives/exhibits/ww2/Pages/services-farm.aspx.

4 Toxic Drift: Pesticides and Health in the Post-World War II South. *Environment & Society Portal* (2013). https://www.environmentandsociety.org/mml/toxic-drift-pesticides-and-health-post-world-war-ii-south.

5 Wills, M. War and Pest Control. *JSTOR Daily* (2018). https://daily.jstor.org/war-and-pest-control/.

6 Pollan, M. *The Omnivore's Dilemma* (Penguin, New York, 2007).

7 Cederberg, C. & Sonesson, U. *Global Food Losses and Food Waste: Extent, Causes and Prevention; Study Conducted for the International Congress Save Food! At Interpack 2011* [16–17 May], *Düsseldorf, Germany* (Food and Agriculture Organization of the United Nations, Rome, 2011).

8 Pathak, V. M., Verma, V. K., Rawat, B. S., Kaur, B., Babu, N., Sharma, A., Dewali, S., Yadav, M., Kumari, R., Singh, S., Mohapatra, A., Pandey, V., Rana, N. & Cunill, J. Current Status of Pesticide Effects on Environment, Human Health and It's

Eco-Friendly Management as Bioremediation: A Comprehensive Review. *Front Microbiol* **13**, 962619 (2022).

9   Brühl, C. A. & Zaller, J. G. Biodiversity Decline as a Consequence of an Inappropriate Environmental Risk Assessment of Pesticides. *Front Environ Sci* **7** (2019). https://www.frontiersin.org/journals/environmental-science/articles/10.3389/fenvs.2019.00177/full.

10 Carrington, D. Plummeting Insect Numbers 'Threaten Collapse of Nature'. *The Guardian* (2019). https://www.theguardian.com/environment/2019/feb/10/plummeting-insect-numbers-threaten-collapse-of-nature.

11 US EPA. Conditional Pesticide Registration (2013). https://www.epa.gov/pesticide-registration/conditional-pesticide-registration.

12 USDA ARS. *Online Magazine*, Vol. 49, No. 1. https://agresearchmag.ars.usda.gov/2001/jan/straw/.

13 US EPA. Phaseout of Class I Ozone-Depleting Substances (2015). https://www.epa.gov/ods-phaseout/phaseout-class-i-ozone-depleting-substances.

14 Overview | Agroecology Knowledge Hub | Food and Agriculture Organization of the United Nations. https://www.fao.org/agroecology/overview/en/.

15 Wallach, J. J., Swindall, L. R. & Wise, M. D. *The Routledge History of American Foodways* (Routledge, New York, 2016). doi:10.4324/9781315871271.

16 Carson, R. *Silent Spring*. (Houghton Mifflin, 1962).

17 Gliessman, S. R. & Alfred E. Heller Professor of Agroecology, UC Santa Cruz | University Library. https://library.ucsc.edu/reg-hist/stephen-r-gliessman-alfred-e-heller-professor-of-agroecology-uc-santa-cruz.

18 FAO. *Hidden Costs of Agrifood Systems and Recent Trends from 2016 to 2023* (FAO, Rome, 2023).

19 Anna Kramer. (December 10, 2014). https://www.oxfamamerica.org/explore/stories/these-10-companies-make-a-lot-of-the-food-we-buy-heres-how-we-made-them-better/.

4

CONSIDERING HOUSING AND
CLIMATE CHANGE

I started this chapter in a kayak on the water,Ventura Harbor, just past the break.The waves are growing as the tide comes in. It smells like pelicans, and I am picking pieces of white Styrofoam from the water. Clear skies, 16 degrees Celsius (61 degrees Fahrenheit), wind nine mph from the southwest.

Housing is about water. It makes sense, if you consider humans are made up of at least 60% water and depend on it for everything we eat and grow.[1] Fresh, available water, but not too much, is a requirement for stable human living conditions. Many cities are located on major waterways critical for transportation, now vulnerable to sea level rise and storms. Before us, generations of humans migrated seasonally to mitigate the impacts of seasonal weather (and water), but now our societies are reliant on static, stable infrastructure.These buildings must stand up to the increasing onslaught of extreme storms and temperatures, sea level rise, drought and saltwater intrusion, and water and air pollution.

In many parts of the world, the ecoregions and local climate are changing. Generally, there seems to be a shift in temperature and condition from the equators toward the poles, where in the future regions will see similar climate conditions as the current conditions in the regions just south of them.

Interviewing experts for this chapter, I really began to understand that the systems we have created around housing are complicated.They are so complicated that they are uniquely different for each country. As a result, some of the insurance regulations and laws in this chapter may be more specific

DOI: 10.4324/9781003376323-4

to American audiences, but will provide a framework for how to think about our challenging housing situation that can be applied to other countries.

A recent paper[2] suggested that the housing market doesn't incorporate the actual cost of the risk from sea level rise into the current pricing structure for most homes. The same underpricing is true for drought and wildfire. They suggest that in America in 2023, climate change risks are underpriced by > 2.4–9.7%, depending on the location and risk type (flooding, sea level rise, fire, etc…).[2] That means that the housing market and people buying houses aren't incorporating the future costs they will have to pay associated with their home's location. They aren't being pushed to consider that they may be buying a house in a floodplain or in an area known for wildfires. This can even mean that entire neighborhoods are abandoned as insurance becomes unaffordable, and dangers from the surrounding ecosystem increase.

Even in the 2020s, many houses are sold with the idea that the climate (and weather) will be the same as now, or as stable as in the 1950s. People take for granted that their beach, island, and ocean cliff houses will remain near the water—not IN the water. Many of the modern ideas about housing are firmly stuck in a post-WWII construction boom that built single-family homes covering huge land parcels, a concept that is neither applicable nor sustainable any longer.[3] There is a real concern that some, even much, of the existing housing stock that was built to house single families may not be livable in the next few decades. Already, insurance companies are refusing to insure houses at high risk from the impacts of climate change. For example, the US state of California is providing insurance to homeowners in high-risk fire areas because insurance companies have already refused to insure these properties. In this way, California is supporting a generous scheme to let people enjoy living amongst the trees and expansive hills. Also, it's completely financially unsustainable in the long run.

Climate change has become one of the largest issues for housing in the past decade. When I spoke with Deana Vidal, she was in San Mateo, the unceded homeland of the Ramaytush Ohlone. It was early March, and it was raining hard. It was raining weekly for over a month. She says that climate preparedness and climate change impacts on housing are issues they discuss almost daily in their offices. When it comes to disaster preparedness, there are a lot of design details that can solve or reduce damage due to storms, flooding, and fire.

For example, she mentions the fires in Sonoma County, California, in 2020.[4] The reason, she says, so many neighborhoods burned is because the fire jumped between single-family house complexes as embers traveled on the wind. These embers got underneath roofs and sparked more fires. However, she says that if the builders had used materials created to reduce the risk of fires, the damage to the neighborhoods would have been considerably less.

Many builders are starting to integrate design choices that incorporate growing climate risk into their builds. There are an increasing number of design details utilizing climate-resilient choices in construction that can reduce the impacts of climate change including flood and fire risk.[5] For example, if an owner or landlord needs to fix their roof, there are small material changes that they can make to prevent future flooding, leaking, or fire problems. In flood zones owners can make sure they have the right water and rain collection systems, or they can move habitable spaces above the level that could flood. Today, developers, architects, and designers are learning about the impacts of climate change and are planning ways to counter them.[6] She assures me that developers and home builders want to get this transition to long-term climate crisis-resistant housing right.

I wonder aloud if we really have to build new homes in light of the climate crisis. There are so many problems with building new houses, including that the materials required to build a house contribute to the emissions behind the climate crisis (lumber, gypsum, even the chemicals), not to mention all the land use issues. The change from wild or agricultural spaces to housing developments in even my lifetime has been incredible, leaving many of the loveliest landscapes unrecognizable. She says that one option is to build back better after disasters. An even better option is to retrofit buildings to keep up with environmental changes. Newer homes are already more up-to-date, but with some planning and work we can make the older homes safer too. Home builders, legislators, and many others in the field are actively working on strategies to mitigate and prevent damage to current housing. What is clear, is that the housing market will continue to face the need for adaptation and to reduce vulnerability to climate shocks as the number of disasters increases.

There are big changes that can be made to our housing as system well.[7] Passive houses, strawbale homes, rammed earth homes, and homes made of other materials that would be called 'non-traditional' are all options. The current concern from the house-building community Deana says, is that these may not be scalable to the neighborhood or city level. The reasons for a lack of scalability are complex but include the level of training that is required for trade smiths (electricians, plumbers, etc.), what is required by the building codes, whether or not it is possible to ensure a non-traditional structure.

It turns out that if you change one traditional material in a building, say from drywall to rammed earth, it functionally changes everything. Thousands of pieces have to come together to build a home in the right sequence. Companies trying to incorporate new climate realities by incorporating non-traditional building materials often spend hundreds of hours negotiating with municipalities just to secure a building permit. This takes a lot of energy and money, but eventually makes it easier to build a more sustainable and resilient structure.

So, what does all of this complexity this mean for a person looking to buy a home?

I don't know that all of the factors that are going to shape the housing market are entirely clear at this point. What is clear is that there are going to be additional considerations for anyone looking to buy a home in the time of climate change. It will be critical to understand the climate risks in your buying area—and how those risks could grow or change over time. For example, if you are near the coast, you need to consider sea level rise, coastal storms, storm surges, and saltwater infiltration into the fresh groundwater. Mountainous regions should consider the influx of high precipitation in short periods with landslides interspersed with drought and fires. Desert

Antarctica, 2013

dwellers should understand the availability of water now and in the future, and the risk of flooding from atmospheric rivers.

There is also a growing concern among both homeowners and the insurance industry that the models used for determining risk to houses and neighborhoods for integrating information are out of date. This could lead to some areas being considered high risk for climate change impacts when they actually aren't and vice versa. It could mean that it is difficult to get

Santa Barbara, 2022

loans or insurance in some areas, creating potential inequity or disenfranchisement in the future. This is an emergent issue that is being referred to as blue lining, and it illustrates why integrating the most accurate models and information on climate risks at every level from the homeowner to the country is important.

Which begs the question—where can I find the best information about regional and housing risks for my area? Many municipalities, cities, and towns now have climate risk assessment tools that residents can use to assess the risks to their homes. In addition to those tools, many countries have government agencies responsible for observing the Earth at different scales and forecasting the risks and impacts for those areas based on climate models. These agencies include NASA,[8] National Oceanic and Atmospheric Administration (NOAA),[9] and Housing and Urban Development (HUD) in the United States. Focusing on different areas of interest and using different tools, these agencies work together to share the best information available with policymakers and the public. In Europe, the European Space Agency (ESA)[10] and regional space branches like ASI[11] and DLR[12] do similar work. In India, the Ministry of Environment, Forest and Climate Change (MoEFCC)[13] does this oversight and planning, and in China it is the Ministry of Ecology and the Environment.[14] Most countries have an agency that researches climate change and provides information to the public. Finding the cutting-edge information, you can trust is crucial to making important financial decisions. Building standards that incorporate climate risks are being developed in the United States and many other countries, but it will be vital to plan and do your research now if you are thinking of buying, or building, a home.

In response to these diverse challenges of the current housing market, as well as the exploding cost of single-family homes, younger generations are making choices that may baffling economists, sociologists, and city planners alike: cohabitation. People are choosing to live together with others outside of their nuclear family structure. Being raised in single-family houses or apartments as latch-key kids with very little adult or peer emotional support left many in younger generations wanting connection in their adult lives. Some still choose to live in nuclear families (~40% of millennials),[15] but many have chosen to live with aging relatives (~25%),[16] and many more live with various cohabiting schemes (~65% of millennials),[17] including cooperatives, artist communities, or with roommates.[15,18] These trends are only expected to continue with younger generations. These communities require less room to house the same number of people than 'single-family' homes, making them more climate resilient, and in many cases, socially resilient.

The idea of cohabitation is obviously not novel. Humans lived together in various community structures since early history. Living apart in nuclear families with single family homes became popular only in the last hundred years. You can see it in the name—it would have been hard to have a 'nuclear' family before the nuclear age.

When I moved to Maine for my PhD, the housing market was deflated and contained some of the oldest houses in the country. The house I lived in was relatively new, and I had wonderful housemates who are still my best friends today. However, in New England, it can get to −31 degree Celsius (−25 Fahrenheit) in the winter, and it is costly to heat the houses. The infrastructure hasn't allowed for electric heating or 'natural' gas to be used in many of the houses, so many places like mine had a huge fuel tank in the basement to heat the house. I always felt like it was equivalent to sitting on a bomb—and I tried very hard not to think about what would happen if my house ever caught fire.

That aside, it was a lovely house that gave me the safety and security to do the hard work of getting my PhD and starting a new life as a scientist. Unable to keep up an empty house when I moved away to work, I sold the house when I graduated. Pandemics, the global financial crisis, and the invasion of many countries later, the housing market prices have more than doubled. I mention this because I know that while I write about the discussions and priorities in the housing market for addressing climate change, many people cannot afford a house—many more than in my parents' time and even fewer than could afford houses ten years ago.

There are intrinsic inequities built into the housing market, including historical redlining, the ballooning cost of student loan repayment debt, the growing crisis of housing stock left empty and used as tax shelters or Airbnbs, leading to the extraordinary prices for houses that were more affordable to purchase only a decade ago.

This is more prevalent in some countries than others, with the World Economic Forum noting that some European countries are culturally very commonly rental societies while others, including Russia and China, are focused on owning homes.[19] Location and demand drive these dynamics, but the availability of clean water, food, and moderate environmental conditions will continue to impact the housing market regardless of a person's renting or owning status.

As the housing market grows in price and complexity, it seems that there may be an increased need for government intervention for consumers. The overlap of areas with moderate climate impacts and affordable housing will continue to shrink, creating a need for innovative solutions. The goals of safe housing for all must continually be prioritized as more people have to move due to climate change impacts, outdated infrastructure, and prices. The uncertainty of climate change may be one of the only governing certainties in the housing market, so planning for the future and educating ourselves about the risks in the area we live is critical. Securing safe, sustainable housing for all may continue to be one of the biggest challenges of our time, and it must be a priority for our governments as well as our friends and families.

While we are considering the risks and challenges, we also have the opportunity to create buildings that will allow us a more sustainable, livable, and happy future. We can incorporate structures and ways of building that are beautiful, and that can play host to plants and animals in novel ways.[20,21] If we are forced to rethink and rebuild our way of living, let us take the opportunity to build better—more consciously and more sustainably. To make something that will last and be appreciated for generations to come.

Things to think about with climate change and housing, in summary:

- Determine what the impacts of climate change will be like in your area, using the best available information. This is often produced by your own government
- Think about whether you want to rent or own, and what the risks or benefits of each might be relative to long-term climate risk in your area
- Think about whether living in a new build that incorporates climate risks or a retrofitted older build is right for you
- Decide if you want to live with others and what the costs or benefits might be as climate disasters unfold
- Take the time to understand how the building has been prepared for climate change, and what the liabilities are (before moving in!)

A FEW RESOURCES FOR FURTHER SUPPORT

Global Climate Dashboard: Tracking Climate Change and Natural Variability Over Time. https://www.climate.gov/.

United Nations Website. Accessed 2024. https://www.un.org/en/global-issues/climate-change.

European Space Agency Website. Accessed 2024. https://climate.esa.int/en/.

Ministry of Ecology and Environment The People's Republic of China Website. https://english.mee.gov.cn/.

REFERENCES

1 How Much of Your Body Is Water? That All Depends. *Discovery*. https://www.discovery.com/science/How-Much-of-Your-Body-Is-Water.

2 Gourevitch, J. D. Kousky, C., Liao, Y., Nolte, C., Pollack, A. B., Porter, J. R. & Weill, J. A. Unpriced Climate Risk and the Potential Consequences of Overvaluation in US Housing Markets. *Nat Clim Chang* **13**, 250–257 (2023).

3 America's Coastal Cities Are a Hidden Time Bomb. *The Atlantic*. https://www.theatlantic.com/science/archive/2023/02/coastal-cities-housing-sea-level-flooding-climate-change/673106/.

4 Glass Fire | CAL FIRE. https://www.fire.ca.gov/incidents/2020/9/27/glass-fire.

5 Wildland Hazards and Building Codes. *OSFM*. https://osfm.fire.ca.gov/what-we-do/code-development-and-analysis/wildland-hazards-and-building-codes.

6 Ignition Resistant Homes. *Wildfire Risk to Communities*. https://wildfirerisk.org/reduce-risk/ignition-resistant-homes/.

7 Environment, U. N. Building Materials and the Climate: Constructing a New Future. *UNEP - UN Environment Programme* (2023). https://www.unep.org/resources/report/building-materials-and-climate-constructing-new-future.

8 Climate Change. *NASA Science*. https://science.nasa.gov/climate-change/.

9 Climate. *National Oceanic and Atmospheric Administration*. https://www.noaa.gov/climate.

10 Home. *ESA Climate Office*. https://climate.esa.int/en/.

11 Homepage. *ASI*. https://www.asi.it/en/.

12 The German Aerospace Center (DLR). https://www.dlr.de/en.

13 Ministry of Environment, Forest and Climate Change. https://moef.gov.in/.

14 Ministry of Ecology and Environment the People's Republic of China. https://english.mee.gov.cn/.

15 Bennett, A. B., Parker, K. & Bennett, J. As Millennials Near 40, They're Approaching Family Life Differently Than Previous Generations. *Pew Research Center* (2020). https://www.pewresearch.org/social-trends/2020/05/27/as-millennials-near-40-theyre-approaching-family-life-differently-than-previous-generations/.

16 Millennials, Retired Parents Living Together (VIDEO). https://www.scrippsnews.com/us-news/housing/millennials-retired-parents-living-together.

17 Donevan, C. Millennials Navigate the Ups and Downs of Cohabitation. *NPR* (2014). https://www.npr.org/2014/11/01/358876955/millennials-navigate-the-ups-and-downs-of-cohabitation

18 Hoffower, H. The Typical American Family Looks Nothing Like It Did in the 1960s, and It's Largely because of How Differently Millennials Are Doing Things. *Business Insider*. https://www.businessinsider.com/millennials-changed-typical-american-family-structure-2019-12.

19 Rent or Buy? These Countries Have the Most Renters vs. Homeowners. *World Economic Forum* (2021). https://www.weforum.org/agenda/2021/04/global-percentage-rent-own-global-property/.

20 UNESCO World Heritage Centre. Works of Antoni Gaudí. *UNESCO World Heritage Centre*. https://whc.unesco.org/en/list/320/.

21 Hundertwasserhaus: Vienna's Colourful & Quirky Housing Complex. *The Culture Map* (2016). https://www.theculturemap.com/hundertwasserhaus-viennas-colourful-housing-complex/.

5

CONSIDERING A FINANCIAL FUTURE DURING CLIMATE CHANGE

Ventura county, 6 am, 6 degrees Celsius (44 degrees Fahrenheit), and clear. No wind and the birds are singing.

Growing up as a young woman, I was always advised to have my own money and my own bank account.

In some regard, money represents freedom. The freedom to leave a situation, to choose your path, to respond to emergencies. I was advised to never rely on anyone else for money.

Recently, I've had students of all ages explain to me that they don't need to save money long-term, because the 'world is ending.' Though sometimes I resonate with that sentiment and have often wondered if putting money away for retirement is superfluous. But this way of thinking is self-defeating. Yes, the physical risks of climate change are growing faster than many people (even scientists) anticipated, but there is still a chance to stabilize climate warming and create a beautiful, sustainable world. It would be awful to wake up 20 years from now, to a world where we have slowed or even reversed climate warming, the planet is healing, and you have no savings. Saving money could also be a useful tool to participate in the fight against climate change. Any excess income can be donated to organizations preserving wild spaces or fighting pollution.

Climate change will continue to underlie everything that we do on this changing planet—and that does include economies and the economic decisions of individuals. How you save money, what industries are dominant, and how the transition to renewable energy occurs is all critically important to

DOI: 10.4324/9781003376323-5

stable economies. Much of the stability of economies built on raw materials, labor, and then finally product distribution may be upended by changes to the availability of materials, or changing demand from consumers like us.

So, this chapter is about money. Many people have strong feelings about money—where it comes from, where it goes, and how it gets there. Myself, I like planning and saving, and I've always been intrigued by the complexity money brings to relationships between people, businesses, and even countries. How we spend our money reflects our values and identities. In some ways it represents our cultures, and there are many guides for how to think about economic valuation and system change. For this chapter, we focus on the financial market changes that are driven by climate change and may impact your finances in the short and long term.

I am not a financial counselor, but as a climate scientist I can see that climate change may affect each part of our lives—including finances. Further, I encourage you to seek our financial help from a financial counselor who understands the evolving investment dynamics within the context of evolving climate change. This could be your bank or employer, and in some cases the local government can provide insights. With all that said, the markets that underlie our global economy currently are, and continually will be, affected by climate change. They will continue to experience changes, both rapid and slow, that will impact nations, corporations, and individuals. I spoke with some of the people in finance who are on the front edge of thinking about these dynamics, to understand what they foresee for the changing market.

Saving money is difficult, but critical. Saving money is also a privilege. It means that you have enough to pay for housing, food, any insurance, all physical needs, anyone else who depends on you and still put money away. This is not the case for many people. In fact, 44% of Americans[1] and 100 million people in the EU don't have enough savings to cover two months of expenses based upon their current income.[2] This is true for countries all over the world—the economic disparities at this time in history rival the days of global oligarchies and are driven by policies that intentionally keep these dynamics in place.

The transition away from fossil fuels will require investment… I don't think Jeff Gitterman[3] was trying to be a finance guy. He studied human cognition in school and is deeply spiritual. He's written books on meditation and finding peace. Jeff is a person who deeply embedded himself in a flawed economic system on purpose. He was looking for a way to improve lives, and he decided a while ago that money was an important way to make a change. He told me on the morning that we talked that "anyone will come to the table to talk about money." Even if it's hard to convince people to change their minds about their morals or politics, you can get people to change if there is an impact to their money.

Jeff lives on the island of Manhattan, on the unceded Lenape land of Mana-hata, where his family has lived for decades. When I speak with him, it's 45 degrees Fahrenheit (7 Celsius) there, and raining, at 9 am local time, with 2′ of snow forecast for the weekend in Upstate. That week turns into the coldest week of the year, unseasonably frigid. When he leans into his screen, and I can see the small Manhattan living room behind him.

Jeff trains others how to invest in financial portfolios that prioritize Earth system thinking and sustainability. He says that the markets are one of the best examples of a complicated system where people must come to the table and compromise. It's hard to do that when there is constant fighting, and often it's even harder to introduce a new dynamic, like investing in climate solutions and sustainability, into the equation. He told me that any type of investing that doesn't incorporate climate change data "is like hanging onto your old map and saying that using the GPS is 'woke' driving."

For investing in stocks, he recommends thematic investing—diversifying a person or company's portfolio to account for changes from climate change. This could include both investing in sustainable stocks, as well as commodities or sectors that need financial support to transition to a more sustainable business model. He says that it's important to realize that the things we must invest in as a society now to ensure long-term resilience may not currently have the lowest CO_2 footprint. For example, it is not currently possible for many if not most companies that are transitioning away from fossil fuels to abandon them completely. While they may be actively working to convert to renewable energy, like wind and solar, in the short term they may continue to emit CO_2 and may not be considered a completely 'green' investment. Financing a transition to renewable energy sources will take time, investment, and capital.

As many universities and businesses are considering divesting from fossil fuel companies through their organization 401k or university portfolios, a challenge is that fossil fuel companies currently own the majority of patents for renewable energy. And divesting from a stock removes your voice as a stock owner. This is specifically important for universities where some of the most cutting-edge thinking could be a force for good in the markets. If a large endowment were to use their voice to drive change (like sharing of patents, for example), the companies would have to consider their requests.

Take geothermal energy as an example. This gorgeous energy source is the main power source in places like Iceland and relies on the underlying geology of the land, transforming the existing heat and energy into power. However, to get to these deep Earth layers that power geothermal you must dig =>1000 feet down, and right now the only businesses with required infrastructure and talent to safely dig so far into the ground are (you guessed it) fossil fuel companies. So, there must be a transition of talent, technology, and resources to make renewable energy viable in the long run from fossil fuel companies and others. So, if we want to encourage technology sharing, some pressure from stock owners may be necessary.

Whether organizations use their financial portfolios to signal the market of a need for change (by withdrawing money) or to use them to be a part of decision making (by owning stock) is a choice for each organization or individual. The goal is the same—to transition to a sustainable, lower-emission economy. I personally do not advocate for spending any more money than we must on fossil fuels or the companies that produce them, but it is an interesting problem we all have to decide how to navigate. One thing we can all do during this transition is awake and aware of how our money is being spent.

Land and water are increasingly in demand but must be preserved for everyone. For example, right now the world requires materials for the energy transition. 'Rare Earth' minerals, lithium, sodium, and graphene, all of which are critical to many zero emission technologies important to a global energy transition.[4] In this scenario, mining companies dig in some of the most remote and important the Earth and claim the proceeds for themselves. Often, if any of the substrate they bring up that is not useful (and may well be toxic) they leave on top of the soil, and it moves into the environment. Since they don't 'own' or have personal investment in the rivers and forests nearby, it's not their issue. This system is not correctly incorporating a number of ecological and social dynamics that need to be interrogated. While the entire concept of taking something from the Earth that is not yours and proclaiming it to be your personal property still mystifies me it is a part of our modern economy and the underlying power of the stock market. This includes everything from corn to bacon to gold and silver.

Many of the global commodities that rely on the current agricultural, energy, and mining technologies will be impacted by climate change. When crops fail or energy grids go down, it impacts the global economy, which like the larger ecosystem, connects us all. In order to protect any money that you or your company/university/bank/town invest, it will be important to think carefully about how businesses are prepared for climate change, and to diversify what you invest in. It is also recommended to save money outside of bitcoin or stocks or real estate, but that is also a luxury and preference.

The global economic system is going to have to adapt. The transitioning economy that will continue to be impacts by climate change shocks may be a perfect time for us to reassess globally whether the dynamics that underpin the global market economy are what we want for the future. The current system negates the social responsibility we all have to each other, and the species around us in many ways. The very rich are buying land and water rights, to ensure against a failing market and resource shocks. This transfer of land from people who have owned it for generations, or lived on it since time immemorial, to the very rich, is not a new dynamic but is a cause for concern. As climate change increasingly introduces shocks to the market, and the world seeks a transition to renewable and sustainable business practices, the global market may have to rethink its structure and function.

Money during a time of climate transition, in summary:

- Saving money is difficult, and critical, for personal security
- The global economic system is going to have to adapt as climate change disrupts processes worldwide
- The transition from fossil fuels will require input and investment to make it sustainable
- Land, water, and minerals are increasingly in demand, but must be preserved for everyone
- Invest in or buy products and services you believe in, but do your research first

A FEW RESOURCES FOR FURTHER SUPPORT

What on Earth Is the Doughnut? https://www.kateraworth.com/doughnut/.

REFERENCES

1 Konish, L. 44% of Americans Can't Pay An Unexpected $1,000 Expense from Savings. 'We're Just Not Wired to Save,' Expert Says. CNBC (2024). https://www.cnbc.com/2024/01/24/many-americans-cannot-pay-for-an-unexpected-1000-expense-heres-why.html.

2 Risking their Health to Pay the Bills: 100 Million Europeans Cannot Afford Two Months Without Income. *Bruegel* | *The Brussels-Based Economic Think Tank* (2021). https://www.bruegel.org/blog-post/risking-their-health-pay-bills-100-million-europeans-cannot-afford-two-months-without.

3 Gitterman, J. *Beyond Success: Redefining the Meaning of Prosperity* (Balboa Press, Bloomington, IN, 2014).

4 Not So "Green" Technology: The Complicated Legacy of Rare Earth Mining. *Harvard International Review* (2021). https://hir.harvard.edu/not-so-green-technology-the-complicated-legacy-of-rare-earth-mining/.

6

CONSIDERING WATER AND CLIMATE CHANGE

I've been waking up with the sun for months. It's early on the West Coast of the United States, and it's still cool despite the ongoing heat wave—the hottest summer recorded to date. Eighteen degrees Celsius (65 degrees Fahrenheit) here, and the birds are singing; there is a mist over the channel.

Water is not a thing. Water is a being. Water is the blood and spirit of this planet, and the story of water is the story of all life. In order to have clean water, you must have a thriving ecosystem.

When we think about rapidly aridifying desert systems, the importance of wetlands to bird habitats, the quality of wine in Southern France, or the delight of mangos from Guatemala—it is the story of water. Our morning coffees, all the food we eat, and our very bodies are all stories of water that has been used, reused, and moved throughout our water planet forever. We live on a blue planet, and clean, freely available water is a critical need for all life. Water underlies everything on the planet. Everything we do, eat, drink, and wear. This is a water planet. Water covers 71% of the planet's surface, with over 97% of that as salt water and ~3% as fresh water. This small proportion of fresh water is crucial for all life on Earth. We humans are made of mostly water and are completely dependent on fresh water to live.

When I give talks about my research, I sometimes tell the story of being trapped on a glacier in an eight-day blizzard snowstorm in the mountains of Alaska near Fairbanks (feet of snow per day! Fieldwork can be hard!). Even this blizzard story is really a story about our relationship with water.

DOI: 10.4324/9781003376323-6

Antarctica, 2013

Antarctica, 2013

When research vessels sail across the Southern Ocean to Antarctica through the Drake Passage—a relentless system of waves that cross the planet and that have taken many lives—that is the story of water.

Nobel Laureate Toni Morrison talks about water in most of her incredible novels. She talks about the 'straightening (of) the Mississippi' and the memory of that river wanting to return to the previous river course—the memories of water, and of all rivers. Even psychics and con artists are renowned for suggesting that missing items are by bodies of water—because anywhere you go on the planet there are always bodies of water nearby. The consistency of water in our lives, the ever presence of water in everything we do, sheds light on why the idea of losing our clean water is so horrific and terrifying.

I remember my first trip to Zurich for a climate summer school I attended during my PhD. The training school was in a small Swiss town, and when I went to Zurich to catch my flight home, I was expecting what I considered normal—a polluted, populated city amidst beautiful countryside. To my absolute shock, that wasn't the case. The river that flows through the middle of Zurich was so clean you could see the fish swimming near the bottom. This brilliant blue water had healthy reeds and all manner of riverine life. Right in the middle of town. It was then that I realized that, as a planet, we have accepted that clean water is a privilege and dirty water a side effect of an industrialized society—but this is neither correct nor true.

The drying of the hottest places on Earth, coupled with the deluge of seemingly randomly timed atmospheric rivers that feel uncontrollable, solidify our feeling as people that the planet is getting hotter, and more chaotic. Flooding events are increasing globally, with temperature increases over the 1.5C global average relative to post-industrial average temperatures are driving more storms that are more than 10% wetter per event. These events cost hundreds of lives per year and the damage from flooding results in losses of around 60bn $US per year.[1] Water is everything to every organism on this planet. If water is polluted, absent, or coming down in buckets, it is no longer our protector. Then it brings destruction and sometimes even death.

There are many climate change issues around water that are now at the forefront of human politics and planning. Some of these issues focus on sea level rise, the overuse of water for agriculture, and heavy rainstorms. Basically, the story of water changing where it is located and how it interacts with our cultures.

There are also growing, compounding problems from growing water pollution reducing our access to clean water. Examples as diverse as the lead poisoning of the small town of Flint, Michigan, in the United States to the rivers in Guatemala outflowing plastic pollution from cities and manufacturing facilities, watersheds are increasingly polluted by industrial chemicals and plastics. These chemicals are often dumped in cities or towns where there is a lack of regulation or where regulation is so relaxed that the watersheds aren't monitored.

Water is cyclical. It will always go toward its source, back from the oceans to the sky to return to the land—but through the course of our recent development we have given it many things to take downstream. Entire forest systems that act as 'water pumps' to circulate and recycle water, making clouds, and wind as the trees breathe.[2] This incredible forest dynamics was discovered by western science only a few years ago, but if there is increased deforestation, then we will lose this incredible filtration and recycling system.

Alaska, 2016

Chemicals, heavy metals, excess nutrients, plastic. Moving across the mountain ranges and deserts of the world, water continues to reflect who we are, and what we are doing with our precious resources. From the acid rain scares of my childhood in the 1980s to the very recent revelation that water carries microplastics in each drop, we have changed the very character of the planet by changing the nature of the water. The next and most crucial question we must ask ourselves now: can we fix it? Can we make the lifeblood of the planet clear again? Or is this relationship permanently damaged?

We know that water is the reason we evolved and the bedrock for all life on this planet. It is intrinsically connected to all life—where stable ecosystems clean and produce water, and clean water allows ecosystems to thrive. When NASA astrobiologists look for life on other planets, they start with looking for water.[3] Many nations, voyaging cultures like the Pacific Islanders and Chinese, fishing cultures like the Chumash and Newfoundlanders, and boat cultures like the Venetians and Dutch, have built their identities, trades, lore, and identities around water. Prairie and desert cultures around the world build cultures around capturing and conserving water. Clean water is life.

And now it comes in a clear, non-disposable plastic bottle.

Clean water is reliant on thriving ecosystems. And all ecosystems on Earth rely on clean water. Just like water is the foundation for all life, clean water relies on natural processes. From the small oysters and mussels who filter out ocean and bay water[4] to the diverse plants in rivers that maintain water clarity, clean water depends on thriving ecosystems. Forested systems, with their intricate belowground mesh of roots, microorganisms, and microbes, are critical to cleaning and purifying water, while they reduce excess runoff, and recharge underground water storage in aquifers. When water is polluted with algae and clouds or becomes stagnant, rivers, lakes, and oceans are in danger. You may have seen it in your local rivers, lakes, or coasts as temperatures rise and ecosystems change.[5] Our water is turning just a bit more green, due to the algae and microorganisms that can thrive in water highly enriched with fertilizers. Polluted water harms the plants and animals that maintain water bodies because the system is an ouroboros, a snake eating its tail, an endless refresh cycle. All creatures are harmed when the water goes bad.

Clean water isn't produced. In my lifetime, water has become a marketed, transferable, bottled, and shipped commodity, in partnership with fossil fuel and plastic companies. When I was a child, I remember laughing at the idea of small plastic bottles of water you could buy; it was such a new idea in the US. Bottling our water hasn't made water more available; it's quite the opposite. Bottled water companies don't create fresh water—they take it from the world's aquifers—deep underground networks of clear water—put it in their bottles and move it far away. They profit off water that isn't theirs. In fact, in most developed countries, the water that comes out of household taps is cleaner than any bottle of water you can buy. Bottled water is about 75% sourced from aquifers and other natural sources, and 25% from tap water, decreasing the ground water available for the local communities. You read that right, companies are bottling city tap water.[6]

For bottled water, there are also fewer regulations that govern how much fecal coliform— poop—can be in each bottle source.[7] Not to mention, bottled water is one of the main ways that microplastics and other toxic chemicals enter the human body.[8] Fighting for the right to clean, accessible water may be one of the biggest battles of future generations. This movement toward privatization (stealing) and transporting water to other locations has the potential to cripple local water systems all over the world. The subsequent input of plastic trash from bottled water into waterways could irrevocably harm other water sources.

So, in summary, bottled water isn't clean, it isn't healthy, and in many cases, it's stolen.

More water, more problems... As climate change progresses, there are increasingly heavy rain events, where too much water floods houses, washes away roads, and captures large amounts of city pollution. The dangerous flood waters that surround houses, cars, and people contain everything from the local roads, buildings, and sewers. They are rife with plastics and sewage and flows directly to critical environments. This influx of trash and waste from floodwaters may harm ecosystems, and hurt the health of the people whose homes it destroys. Our inability to capture and use the water that rains on our cities and towns makes it all the more dangerous—leaving us unprepared for the following periods of drought. This shift in how water moves through the Earth system has felt sudden, with increasing water-driven disasters in less than ten years,[9] and with flooding or unseasonal impacting all parts of the globe (even Antarctica).[10] It is the result of climate change.

But water is life. And our relationship to it must be mended. The water towers of the world-high mountain glaciers and ice sheets are being lost at incredible, un-forecasted rates. The combined loss of water storage and water quality due to ecosystem degradation could affect all life on Earth. Knowing that we are on an unsustainable path and that water pollution is expanding logarithmically, we have the opportunity to heal our relationship to water. Instead of planting crops in places where the aquifers are running dry, letting our waterways in cities fill with plastic trash, and letting development remove the critical plants and animals that protect rivers, lakes, and estuaries, we must begin to work together with the ecosystems to honor and enhance the natural processes that clean water. There is no Brita filter to clean water as strong as the natural world. The only solution to a crisis of clean water is to protect and restore our critical ecosystems—and to allow the water cycle the space and biodiversity it needs.

For more information, I spoke with Jennifer Holm[11] in Anchorage, Alaska, the unceded home of the Dena'ina people. She is a researcher at the Lawrence Berkeley National Laboratory, but also thinks a lot about nature-based solutions.[12]

She says that nature-based solutions are technologies we can use now—that we already know work—because we have had biology working on our side for centuries and have seen it in action all around us. There are three essential strategies that The Nature Conservancy and others have identified as critical for enhancing our relationship with the natural world.[13,14] They have identified:

Alaska, above the Arctic circle, 2023

Jarvis Glacier, Alaska, 2016

1   *Protect the ecosystems that we still have*—and this represents the biggest impact and least cost. It's a win-win for everyone nearby and all of us who are dependent on clean water and air.

2   *Manage ecosystem services* and enhance them, making them more efficient, or more sustainable. Improve and sustain efforts to engage with the natural environment and keep ecosystem services functioning to make sure we are adequately managing our interactions with the life around us.

3   *Restore the environments that we have already degraded.* This is a mission critical for continuing to live on a planet that produces all of our food, water, clothing, air, and everything else you could possibly mention. Restoring an ecosystem takes the most time and costs the most money—but often strategies like reforestation, revegetation, and replanting native species are the least expensive options for restoration and have the greatest impact.

Reforestation and coastal regeneration work best with land that has historically been, or has seen forests, kelp beds, mangroves, or any other vegetation type because the land remembers and can more easily revert back to its previous state. Though it can sometimes be hard to figure out the best places for restoration—making sure to plant the right trees in the right places, and thus enhance the portals for wildlife, water retention, and purification while not exponentially increasing wildfire risk or pest outbreaks, but even one individual can plant a tree and make a difference. There are also companies that professionally plants trees in the correct locations, helping to make sure trees fully survive, that are people-centered, helping the livelihoods of communities, which is very cool and not expensive to support. Restoration can have so many benefits including absorbing the excess CO_2 we are emitting into the atmosphere, cooling from shade (can help with urban heat!), stabilizing the local climate and water cycle, improved air quality, increasing biodiversity with homes for birds and small creatures, and improving our mental health by knowing that there are strong ecosystems nearby (this is real and based on research).[15,16] To maintain a strong water cycle and clean the water that we have degraded, we will require ecosystem restoration, protection, and ongoing management.

Despite these advances, the problem with water pollution will never be a problem of how to build better nets to capture pollution. Instead, we must prevent the plastic, oil, sewage, and all other pollutants from ending up in the watersheds at all. Clean water must be considered a right for all and protected as such. If we must utilize materials we know to be toxic, we must keep them separate from the ecosystems and clean water of the world. It seems like a big task, but that is a decision we must all make together—how we wish to preserve life on this planet. Our blue planet.

Water for life, in summary:

- Clean water is reliant on thriving biodiversity (and vice versa)
- Water isn't produced by people or companies, but a birthright of all life

- In the time of climate change, more punctuated water events may cause more disasters
- But water is, and will always be, crucial to all life on the planet, and critically important to protect

A FEW RESOURCES FOR FURTHER SUPPORT

Centers for Disease Control and Prevention. https://www.cdc.gov/healthywater/emergency/making-water-safe.html.

Guest Post: The Spirituality of Water. https://potomac.org/blog/2015/6/5/guest-post-spirituality-water.

Why Is Water So Important for Life As We Know It? https://astrobiology.nasa.gov/education/alp/water-so-important-for-life/#:~:text=Big%20Ideas%3A%20All%20living%20things%20need%20water.,thread%20between%20all%20living%20things.

REFERENCES

1 IPCC. Climate Change 2021: The Physical Science Basis. Contribution of Working Group I to the Sixth Assessment Report of the Intergovernmental Panel on Climate Change. In *IPCC ARG6* (eds. Masson-Delmotte, V. et al.) 1–2391 (Cambridge University Press, 2021). doi:10.1017/9781009157896.

2 Aragão, L. The Rainforest's Water Pump. *Nature* **489**, 217–218 (2012). https://doi.org/10.1038/nature11485.

3 Searching for Water in the Solar System and Beyond. *NASA Science*. https://science.nasa.gov/missions/hubble/searching-for-water-in-the-solar-system-and-beyond/.

4 Bringing the Bay Home: Outdoor Education, Oyster Filtration, and a Backyard Report Card. https://www.cbf.org/blogs/save-the-bay/2020/09/bringing-the-bay-home-outdoor-education-oyster-filtration.html.

5 US EPA. The Effects: Dead Zones and Harmful Algal Blooms (2013). https://www.epa.gov/nutrientpollution/effects-dead-zones-and-harmful-algal-blooms.

6 Understanding Bottled Water. https://extension.psu.edu/understanding-bottled-water.

7 Ahmed, W., Yusuf, R., Hasan, I., Ashraf, W., Goonetilleke, A., Toze S. & Gardner T. Fecal Indicators and Bacterial Pathogens in Bottled Water from Dhaka, Bangladesh. *Braz J Microbiol* **44**, 97–103 (2013). https://pmc.ncbi.nlm.nih.gov/articles/PMC3804183/.

8 Plastic Particles in Bottled Water. *National Institutes of Health* (NIH) (2024). https://www.nih.gov/news-events/nih-research-matters/plastic-particles-bottled-water.

9 2010–2019: A Landmark Decade of U.S. Billion-Dollar Weather and Climate Disasters. *NOAA Climate.gov*. https://www.climate.gov/news-features/blogs/beyond-data/2010-2019-landmark-decade-us-billion-dollar-weather-and-climate.

10 Warming in Antarctica | Center for Science Education. https://scied.ucar.edu/learning-zone/climate-change-impacts/warming-antarctica.

11 Jennifer Holm Profile | Lawrence Berkeley National Lab. https://profiles.lbl.gov/21921-jennifer-holm.

12 Nature-based Solutions. IUCN. https://iucn.org/our-work/nature-based-solutions.

13 Cook-Patton, S. C., Drever, C.R., Griscom, B.W., Hamrick, K., Hardman, H., Kroeger, T., Pacheco, P., Raghav, S., Stevenson, M., Webb, C. & Yeo, S. Protect, Manage and Then Restore Lands for Climate Mitigation. *Nature Climate Change* **11**, 1027–1034 (2021).

14 Leavitt, S. M., Cook-Patton, S. C., Marx, L., Drever, C. R., Denney, V. C., Kroeger, T., Navarrete, D., Nan, Z., Novita, N., Malik, A., Pelletier, K., Hamrick, K., Granziera, B., Zganjar, C., Gonzalez, J., Ellis, P., Verdieck, J., Ordóñez, M. F., Gongora, C. & Del Castillo Plata, J. *Natural Climate Solutions Handbook: A Technical Guide for Assessing Nature Based Mitigation Opportunities in Countries* (The Nature Conservancy, Arlington, VA, 2021).

15 Nurtured by Nature. https://www.apa.org https://www.apa.org/monitor/2020/04/nurtured-nature.

16 Capaldi, C. A., Passmore, H.-A., Nisbet, E. K., Zelenski, J. M. & Dopko, R. L. Flourishing in Nature: A Review of the Benefits of Connecting with Nature and its Application as a Wellbeing Intervention. *Int J Wellbeing* **5** (2015).

7

CONSIDERING HAVING CHILDREN
DURING CLIMATE CHANGE

9:10 am, Paradise Valley desert outside Los Angeles, near Palm Springs. Clear, sunny, and low 60s. It snowed on the surrounding peaks last night and the wind has been consistent throughout the night.

When I talk about climate change with people in their late teens and twenties, they are distraught about the question of children. They often ask me if they should have children. The real question they are asking is, "Is it worth it to have kids? Will the planet be habitable enough for me to give a child a good life?" In 2023, when I started this chapter, numerous ecosystems were already on the verge of collapse. As I finish it in 2025, I am constantly thinking about whether we have passed critical ecosystem tipping points across the world—with unknown negative consequences. That makes me very concerned about what the future will look like for children born in the decades after this writing.

Having children is, and was, a very personal decision. The ability to make this choice and even talk about it as a choice is, unhappily, still a privilege.

There are over 8 billion people on the planet in 2024 and the UN estimates over 150 million children who are orphaned. That number is likely an underestimate. Both of those facts make me worried about the quality of life of any child I would have. I am uncertain if the planet will be a wonderful place to live when they grow up, as the future we will have is completely dependent on the choices we are making as a global community right now.

I was going to have kids and a farm by my late 20s. That was the plan. I really love children and always say that if the planet were not in crisis, and

DOI: 10.4324/9781003376323-7

I didn't feel the need to work in climate science, I would be a Kindergarten teacher. As it is, the planet needs advocacy and is full of humans, and so far, I have decided not to follow through on my dream of children.

It was hard to write this chapter because it is such a personal, gut-wrenching question: whether to have kids during a time of global climate change. Especially with so many families unsure of the safety and security of those children in the future or at present. Especially with the reproductive rights of women threatened globally, and in my own country. So, I turned to experts to help guide us through thinking about this very real decision.

I spoke with reproductive expert LC De Shay at their home 2hrs south of Amsterdam. They have 18 years of experience in the field of reproduction and advocacy and are finishing their graduate degrees. It's winter, and dark at 6 pm there, sleeting and icy, with some snow. LC is checking on the progress of the traffic and the return of their partner to the warm, quiet home where elementary and preschoolers watch German language cartoons. LC is a genderqueer Black American living abroad. Their background includes Southern Delta, Creole, and Appalachian ancestry, and relations to the Choctaw and Mvskske. The African side of their family has been in the US since 1736. They know about their heritage—an effort of love and dedication to follow handwritten records of parents and parents' parents back hundreds of years.

LC answers the question I ask about whether climate change stops us from having children. "Not to be dismissive, but this is not the first time in human history we all thought our species was going to die." They say that they struggle with the need to protect younger generations, and a sense of failure that we are all facing the question of whether to have children in times of climate change. This suggests that the climate crisis is causing many to examine their plans for children—and their ability to give those children a world where they can flourish. For each of us to answer that question we have to consider many aspects of that choice, including, but not limited to, climate change.

In a just world, having children is a choice. LC says the first thing that a person must consider when thinking about having children, is that for women, the decision to have children is a decision that happens every second of every day after puberty. If your identity drives your social life and you are sexually active with others who can have children—you increase your chances of having children in the future and need to make choices about protection and future family.

Every opportunity that one takes or doesn't take to be physically intimate with someone who could begin a pregnancy is either moving forward or away from children. It reflects your values, as it may lead to pregnancy.

LC says that anyone can make a declaration about whether they want to have children, but the question ultimately, is what are your daily choices? Many people will decide what they do and don't want to happen in their lives at one point in time, but their values and daily habits are creating their everyday lives. Ultimately, it doesn't matter what you say you will do on paper, what matters the most are your daily choices. It is critical to develop a manageable plan for your intimacy that aligns with the actual way you live your life.* If we are concerned about the world our children will inhabit, our decisions about how we address climate change, and our sexual health decisions, should reflect that.

But even in planning for children these days, you also need to plan for an unstable future in a warming world. The changes in the global climate necessitate a new series of questions. Questions like: What is your support network like in times of hardship? What will the weather, food, water, and seasons look like in the next 5, 10, 20 years? Are you affluent, or do you reside in a country that is affluent enough to provide sufficient pediatric care? What social services or policies exist in your country to support your plans to have children? What existing challenges in your life do you have that will become worse because of climate change?

Family planning must now include the environmental factors affecting our lives in addition to our personal goals and directions. Adding climate change to the conversation about family planning makes these decisions infinitely more complicated.

Planning for a new climate future is critical to parenthood. It's hard to plan for an uncertain future, and there is an underlying question of social status and location. As a parent you will have to now also be hyperaware of the changing environmental situation. In the time of climate change, having children responsibly will require access to climate and environmental information, and the ability to understand that information. Those things are simple for the privileged but difficult for many others. Often the inequities of reproductive rights directly reflect the inequities of climate change, meaning, the people who are most affected by poor access to medical care and educational opportunities, are the same people most affected by climate change.

Biologically, getting the right support for people who have children isn't about getting the right positive affirmations every morning—it's medical care and attention. It's about healthy community support and equitable policies that incorporate climate risks. Perfectly, LC says, the answer to both problems of inequity in climate change and reproductive rights is collective advocacy. Lifting people up and enabling choices through education and resource sharing will mitigate two challenges at once.** But there are also gaps in understanding what hardships the most vulnerable people among us will face—because the daily situations for those most affected by climate are worsening faster than the funding for research on climate change or women's health is growing.

Consider climate, and your role in it. The international news has been alive recently with stories of the oncoming 'baby bust.' The birth rates globally are slowing for the first time since just after World War II, and there is both excitement and concern about what that could mean. Within the context of emissions and climate change, we know that each individual leaves a carbon footprint in their lives—meaning that the number of people on the planet directly influences our ability to stabilize the climate and slow emissions.

The 20th century saw an increase in the human population that is almost difficult to comprehend. In the 1900s global populations were 1.6 billion in 1900, increasing to 3 billion in 1960. It then increased non-linearly to 5.3 billion in 1990 and is now around 7.8 billion. That is a ton of people. During that same time, however, the birth rate increased and then steadily declined. For example, in 1960, on average women had just over four children, down to three children in 1990 and only just over 2 by 2019. Since 2019 the birth rate has decreased to the point that it has surprised many experts—down to two children on average per woman in 2024. In some countries the women had a few more children than two, and in many, they had fewer than two.[2] Whether it's the cost of living, the cost of children, a concern for the state of the environment, a worry for the state of the world, or access to birth control, medicine, and education, many people are independently deciding to have fewer—or no—children.

Children have also recently been considered the largest negative contribution that an individual can make to climate change. One child can add up to 9,441 metric tons of CO_2 to the environment per year.[1] That's more than all the driving, flying, and eating unsustainably a person could do, and then multiply by five. This is obviously not caused by the child themselves, but by the consumption and carbon use of many children (and their families), globally. Adding to the difficulty of deciding whether to have children is the knowledge that the carbon footprint of each new individual on the planet is enormous.

The choice to have children is (ideally) entirely personal. I hope for a world where every child is a choice—that the responsibilities of having children aren't forced on anyone. We are not in that world yet. The world we live in often minimizes women's choice—even in light of all of the challenges facing every new generation—and is already rolling back many freedoms afforded to even our mothers. As this struggle for equity and equality continues, I think often about how to strongly I support education, choice, and medicine for all people with wombs. Societal resilience through health and education is also climate resilience.

In my own decision-making, I do often hope that there will always be children in my life. That I could be a good enough person to be a steady guiding light in times of hardship and growth. Writing these words have been an important personal reflection for me as well, as I remind us all to continue toward climate action that align with our values. This includes the huge decision about whether to have children or not.

Thinking about babies, in summary:

- In a just world, having children is a choice made by the child-bearer
- Planning for a new climate future is critical to parenthood
- Consider climate futures, and your role in creating the future

NOTES

* For people without easy access to medical care who need support during a pregnancy, there are already often on-the-ground resources that are not adequately advertised. For example, the World Health Organization has mobile units in many countries, but people may not know about it. You can find United Nations, country-wide, or regional resources online or by telephone. There is care available. It may just not always be easy to access geographically.

In some cases, they can either come to you or help you get to them. This is not sufficient, but it is better than nothing. Community resources are also available on the Internet. A recent story about a part of Mississippi in the United States where there was no childcare provider for over 150 miles came to the attention of people on Twitter. Within 18 months, people had launched mobile clinics (which is better for hurricane season), and there is now a resource provider for the state—Momnibus. Providers are working through all the gender and healthcare challenges worldwide, but often, there must be a request or call for advocacy.

A FEW RESOURCES FOR FURTHER SUPPORT

Make Mothers Matter at the United Nations. https://makemothersmatter.org/delegations/un/.

UNICEF Model Mothers Lead the Way for Moms and Their Babies. https://www.unicefusa.org/stories/unicef-model-mothers-lead-way-moms-and-their-babies.

The Baby-Bust Economy: How Declining Birth Rates will Change the World. https://www.economist.com/weeklyedition/2023-06-03.

Should Climate Change Keep You from Having Kids? https://yaleclimateconnections.org/2024/02/should-climate-change-keep-you-from-having-kids/.

Climate Mitigation Gap: Education and Government Recommendations Mss the Most Effective Individual Actions, Seth Wynes and Kimberly A. Nicholas Published 12 July 2017 • © 2017 IOP Publishing Ltd Environmental Research Letters, Volume 12, Number 7 Citation Seth Wynes and Kimberly A Nicholas 2017 *Environmental Research Letters* 12 074024 DOI 10.1088/1748-9326/aa7541.

REFERENCES

1 Wynes, S. & Nicholas, K. A. The Climate Mitigation Gap: Education and Government Recommendations Miss the Most Effective Individual Actions. *Environmental Research Letters* 12 074024 (2017). https://iopscience.iop.org/article/10.1088/1748-9326/aa7541.

2 Pretty, J. *The Low-Carbon Good Life.* (Routledge, 2022). https://doi.org/10.4324/9781003346944.

8

CONSIDERING POLLUTION
AND CLIMATE CHANGE

Cloudy and muggy. Fifteen degrees Celsius (60 degrees Fahrenheit). The entire US is breaking heat records and by the coast the cool Pacific is meeting these temperatures with evaporation and a marine layer that has stayed for four weeks.

A good friend of mine texted the other night and asked if she should check the soil near her house for PFAS. She recently had a child and was worried her baby might eat some of the dirt outside while playing. She said, "should I get it tested for PFAS (Per- and polyfluoroalkyl substances), or is it not even worth it since this stuff is everywhere now?" PFAS were created in the 1940s and are known as 'forever chemicals' because they don't break down in the environment or through most industrial processes. In 2019, our research team found them on the top of Mount Everest, and other research teams have identified them at the bottom of the Marianas Trench—the deepest part of the global ocean.

There is a certain amount of risk that we each accept in our daily lives. If we drive a car, ride a bike on busy roads, or eat junk food, we know some of these behaviors may have negative consequences and we accept the risk. We chose to participate anyway. Pollution isn't like that, because we often don't have a choice or a say in the risk that pollution presents in our lives, or our ecosystems. At present, we know that plastic pollution is ubiquitous, and we are still discovering the huge footprint of these substances on our world. Pollution affects the water in homes all around the planet, pesticides, and plastics are in our food, water, blood, and the air we breathe.

DOI: 10.4324/9781003376323-8

When households started using Teflon, it was marketed as a safe, non-stick cooking solution. Companies had invented Teflon to coat pans and other kitchen items so that eggs and other sundries slide right off after cooking. It was a big upgrade from cast iron skillets at the time. However, as Teflon moved into the watershed and throughout the local ecosystem, it caused stillbirths, birth defects, cancer, and animal deaths, and the company that made it was called into court to argue that their non-stick pans were safe for use. They were not safe, there were significant known health impacts across all types and ages of humans, but despite this we still use Teflon and similar products today.

My biochemical mentor in my PhD told me that every time society bans a chemical because of the damage they do to the ecosystem or human health, ten more sprout up in its place to do the same job. None of these new ten have been thoroughly tested for their affects in the environment or in human bodies. Eventually, these too will be banned if there are enough cases of serious side effects, and the cycle will begin again. This can lead to the exponential growth of chemicals in our lives, and in the environment. Since the advent and use of DDT, a pesticide widely sprayed in the 1950s and 60s to kill mosquitos, at least ten of these new or replacement chemicals have sprung up in each use class.[1] At the present moment, over 160 million chemicals have been created for human use.[1a] This includes pesticides, flame retardants, plastic polymers, microplastics, cleaning solutions, fertilizers, polyester for clothing, and hundreds more types and qualities across the markets. That's 60 years of new chemicals of every class and type, every year. Thousands and thousands of them released into the environment and absorbed into our bodies.

So, I told my friend it wasn't really worth testing the soil. They don't live near any chemical plants, manufacturers or and high use areas, so the residual chemicals they find in the soil will be a similar concentration to anywhere else. Similar to what is probably already in their bodies.

And there is a lot in our bodies.[2] Recent studies have found microplastics used for textiles, Tupperware, and tires in breast milk,[3] placentas,[4] and semen.[5] They are in our blood, our hair, our lungs, everything we eat and drink, and the air we breathe. They strangle sea lions and starve seabirds to death. Half of the dust in your house may be made up of plastic particles thinner than a strand of hair. It's disgusting and horrifying. And completely unregulated.

Plastic is also not recyclable. I have gotten into many fights about this on the internet and with family members—but it is true. Lawsuits from States and countries are starting to counter this myth in 2024, showing that oil companies drove the myth of plastic recycling as a way to increase profits. Less than 5% of post-consumer use plastic is recycled and turned into something new. Less than 10% of all plastic ever created has been recycled.[6] That means that since the 1940s, every knick knack, every Tupperware, every rotary phone, every plastic fork you used once, every straw, every cheap sweater, every phone case, every plastic number tag from a restaurant, every sofa cover, every piece of plastic (except for 10%) still exists in its original form in a landfill somewhere.

Consider that. A beautiful blue sweater that is 50% cotton and 50% polyester weave sheds a small ball of fabric by the seaside. The ball will be carried by the breeze into the ocean, where it will travel on the currents and slowly make its way down to the bottom of the ocean to settle in the sediments. Or it will be eaten by a sea bird and slowly start to fill its stomach with undigestible substances until they pass that on to their chicks, or they starve to death because there is no room for food in their stomachs. Now multiply that ball times every garment you own, and then time 8 billion people. Plastic is everywhere.[7-9]

The combined issues of climate change and pollution are on my mind every day. But there are new companies and visionary people working to clean ecosystems from stark plastic pollution, right now. I have been following The Ocean Cleanup project since Boyan Slat[14] started publicly suggesting cleaning the Pacific Ocean garbage patch with huge nets as a pre-teenager. The idea that we could use these specifically developed, technologically advanced, giant net systems to remove the plastic drifting in the Pacific Ocean (twice the size of Texas everyone says! Like I understand intuitively how big Texas is. Big, I guess) was novel. Since that time, and despite MANY loud detractors, they have created a safe and successful removal system that, as I write, is removing all of the trash from the Pacific Ocean gyre. They also placed beta version trash-capture net systems at the mouths of crucial waterways around the world. The netting system in Guatemala is constantly in the news because of how much trash it captures, and I have seen the one in the Los Angeles harbor in person. They are extraordinary.

The very bad news is these plastics and forever chemicals aren't alone. There are mines that seep mercury, lead, and other heavy metals into the nearby streams, killing animals and making people very sick.[10,11] These metals are critical for the functioning of everything from electric vehicles and Iphones, so the horrifying pollution from these mines are considered a necessity, a necessary evil. The notion that electrification is without

Banff, Canada, 2014

consequences and is a requirement for a 'cleaner' energy future is therefore not without problems. There are companies that are claiming to do 'ethical mining' in Canada and other parts of the world with rare Earth minerals—but I do wonder if the First Nations whose land they have taken feel that this mining is actually ethical. I am always struck by the number of materials that are within each trashed plane, car, and battery—and reusing these materials would be the first order choice for a truly clean energy transition. Especially as the fossil fuel companies are losing public support for the production of oil and gas for engines, and they move toward increasing plastic production.

I have written a lot about the 'pristine' Arctic, the northern lands far away from 99% of the world's populations, and the hidden but very real footprint that humans have left on those lands. The testing of nuclear bombs in the frozen soil and oceans, dumping everything from nuclear waste to pesticides, and the mining I have discussed are all becoming increasingly frightening as the Arctic ice and permafrost melts and the water moves these toxins into the oceans and air.[13] I have traveled the world talking about this and am constantly working to get a better idea of the how, when, and where these chemicals enter and are circulating throughout our environment, food and water, and our bodies. It is these exacerbating pollution effects that worry me the most.

This seemingly dire, intractable problem of pollution compounding the damage of climate change, however, has a very positive lining. Most of the problematic plastic, forever chemicals and rare earth metal goods are for consumer use. They are the plastic bottles that hold coke and the lining for potato chip packages. Meaning, we have choices in what we buy. It is probably next to impossible to completely phase out plastic and forever chemicals from our daily lives, but we can make choices to slow the problem.

For example, recent plastic bag bans in many countries have reduced use by as much as 80%, with a subsequent 60% decrease in plastic entering the waterways.[12] Plastic straw and cutlery bans have resulted in similar successes. Movements to slow the use of 'fast fashion' where big box stores use unethical cheap labor to produce mostly plastic derived clothing, have also met some success. The reuse of clothing from thrift stores or rental stores is increasingly popular.

Plastic pollution and chemical use in our daily lives is increasingly something that we can control. We are totally empowered to make choices that can make a big difference in our health and safety, and in the pollution problem overall.

To fight pollution in our daily lives, the United Nations recommends the framework Reduce and Reuse, Recycle, Reorient. Reuse is pretty straightforward—the idea is to use items as many times as possible, to keep them out of landfills and to reduce the need for more of the same item to be manufactured. Recycle can be accomplished with new technologies but will always be a second place to reducing and reusing—especially since plastic recycling technology has a long way to go to be a fully recyclable commodity. The UN also recommends removing fossil fuels subsidies, and enforcing design guidelines that make plastic products more recyclable. These are things you can write your government representatives to encourage action. Reorient is a novel way to say, use something besides plastic. From bringing your own grocery bags and reusable mugs to stores, to using washcloths to remove makeup, there are tons of sustainable alternatives that use less plastic and fewer chemicals. The pollution crises may feel overwhelming, but it is one of the most promising areas in which individuals can make a real difference, right now.

Pollutants in the world, in summary:

- Plastic and chemical manufacturing has increased exponentially since the 1940s

- While some items may be critical, minimizing household use of chemicals and plastics can make a big difference
- Focusing on Reducing and Reusing, Recycling, Reorienting are excellent first steps toward a more sustainable future
- Your elected representatives have the most power to decide what the environmental pollution of the world looks like—call, email, write, and visit them

A FEW RESOURCES FOR FURTHER SUPPORT

The impact of plastic on climate change. https://www.colorado.edu/ecenter/2023/12/15/impact-plastic-climate-change.

REFERENCES

1 Guidance on Chemicals and Health. https://www.who.int/tools/compendium-on-health-and-environment/chemicals.
2 Microplastics Are Inside Us All. What Does that Mean for Our Health? *AAMC*. https://www.aamc.org/news/microplastics-are-inside-us-all-what-does-mean-our-health.
3 Carrington, D. Microplastics Found in Human Breast Milk for the First Time. *The Guardian* (2022). https://www.theguardian.com/environment/2022/oct/07/microplastics-human-breast-milk-first-time.
4 Ragusa, A., Svelato, A., Santacroce, C., Catalano, P., Notarstefano, V., Carnevali, O., Papa, F., Rongioletti, M. C. A., Baiocco, F., Draghi, S. & D'Amore, E. Plasticenta: First Evidence of Microplastics in Human Placenta. *Environment International* **146**, 106274 (2021).
5 Montano, L., Giorgini, E., Notarstefano, V., Notari, T., Ricciardi, M., Piscopo, M. & Motta, O. Raman Microspectroscopy Evidence of Microplastics in Human Semen. *Science of The Total Environment* **901**, 165922 (2023).
6 Sullivan, L. The Myth of Plastic Recycling. NPR (2022). https://www.npr.org/2022/12/08/1141601301/the-myth-of-plastic-recycling.
7 Aves, A. R., Revell, L. E., Gaw, S., Ruffell, H., Schuddeboom, A., Wotherspoon, N. E., LaRue, M. & McDonald, A. J. First Evidence of Microplastics in Antarctic Snow. *Cryosphere* **16**, 2127–2145 (2022).
8 Kvale, K., Prowe, A. E. F., Chien, C. T., Landolfi, A. & Oschlies, A. The Global Biological Microplastic Particle Sink. *Scientific Reports* **10**, 1–13 (2020).
9 Primpke, S., Beyer, B., Gütermann, J., Katlein, C., Krumpen, T., Bergmann, M., Hehemann, L. & Gerdts, G. Arctic Sea Ice Is an Important Temporal Sink and Means of Transport for Microplastic. *Nature Communications* **9**, 1505 (2018).

10 Dana, L. P. & Anderson, R. B. Mining and Communities in the Arctic: Lessons from Baker Lake, Canada. *International Journal of Entrepreneurship and Small Business* **22**, 343–361 (2014).

11 Tang, L. & Werner, T. T. Global Mining Footprint Mapped from High-Resolution Satellite Imagery. *Communications Earth & Environment* **4**, 1–12 (2023).

12 Plastic Bag Bans in the US Reduced Plastic Bag Use by Billions, Study Finds. *World Economic Forum* (2024). https://www.weforum.org/agenda/2024/01/plastic-bag-bans-reduce-waste/.

13 Miner, K. R., D'Andrilli, J., Mackelprang, R., Edwards, A., Malaska, M. J., Waldrop, M. P. & Miller, C. E. Emergent Biogeochemical Risks from Arctic Permafrost Degradation. *Nature Climate Change* **11**, 809–819 (2021). https://doi.org/10.1038/s41558-021-01162-y.

14 Boyan Slat | Founder and CEO of the Ocean Cleanup. *The Ocean Cleanup* (2014). https://theoceancleanup.com/boyan-slat/.

9

JOINING THE FIGHT AGAINST
CLIMATE CHANGE

San Diego California. Twenty degrees Celsius (68 degrees Fahrenheit) and muggy. The fog layer from the evening is burning off, I can hear the ocean, and the cushion beneath me is still damp. There are passenger planes taking off nearby, and they pierce the ambiance.

So, the question I always receive—no matter where I am talking or with what audiences—is what can I do about climate change?

In some ways it's like asking what I can do about an avalanche—after it is halfway down the mountain—not much. However, each one of us can take actions—both big and small—so that this climate change 'avalanche' doesn't get worse. We have a window of opportunity right now to decide what our future looks like, together. One person cannot force the oil companies to stop increasing drilling and emissions, and one person cannot alter the fabric of our buy and dispose society. One person, on their own, can do a LOT to limit climate change risks, restore their local ecosystems, and limit the pollution they create. I hope that the previous chapters have outlined some of the small and large steps you can take right now, today, to make a difference.

The action that is required to slow climate change is going to require a collective decision by all of us—and I mean all of us— to stop polluting the atmosphere with greenhouse gases. This is going to mean stopping leaks of methane that emit from landfills and pipelines, changing the way we create—and use!—concrete—and covering less of our beautiful, blue world with it. It will mean using fewer fossil fuels for cars, airplanes (especially private planes), factories, power plants, you name it. Stopping deforestation

DOI: 10.4324/9781003376323-9

Banff, Canada, 2014

and preserving 30% or more (!) of the planet for wilderness. That is not something that one person can do. But all of us working together—this is exactly the kind of collective work and change-making that our governments were made for.

So, for these big, cooperative actions, we have to work together to convince our global governments to act. That's why the problem sometimes seems so intractable. Because convincing an entity as amorphous as a whole government (or all the world governments even) to act seems desperately hard. There are lots of folks leading direct actions, teach-ins, and campaigns to make this change happen. You, as an individual, can seek them out and support them (list at the end of this chapter).

So, what can you do as an INDIVIDUAL? There are multiple, overlapping crises that we are facing. The climate crisis is one, but the biodiversity crisis and the pollution crisis are also happening right now. In biodiversity, the planet is facing the sixth mass extinction—brought on by humans. That means that up to 38% of the animals and plants on the planet are struggling right now.

So, what can you do to help with the biodiversity crisis? As discussed, you can create a habitat. Restore wildlands or pollinator spaces in your neighborhoods. If you have a lawn, stop tending grass (and please, please, please don't fertilize or poison it) and plant pollinator plants, bushes, or trees. If you have a patio, plant flowers in a small patio box and provide food and clean water for the local birds or butterflies. if you don't have a lawn or a patio— use your local parks and advocate for less grass* that requires huge amounts of fertilizer and water and more trees, pollinator flowers, and shrubs.

This monoculture creates a biodiversity desert that is devoid of life—and laying poison on grass is a huge source of pollinator death).

There are also a few outstanding efforts that you can be a part of by international governments and nations to make a big difference in the biodiversity crisis. These include removing streets to replace them with trees, cleaning beaches and waterways, and advocating for regulations to limit pollution. Small gestures like planting wildflowers and trees on highway on-ramps and barriers, reforesting parks, and planting new avenues of trees in cities—these little things can make a huge difference. The 30 by 30 project—where nations (regions, cities, and towns) have committed to preserving 30% of their natural area as wilderness by 2030 would make an incredible difference to all animal species and a great diversity of plants. You can encourage your local region to commit to preserving 30% (at least!) of the connected, undeveloped land in your area, expanding the project and making a huge difference with your advocacy.

Creating nature areas of refuge in your yards, patios, and cities may sound inconsequential, but as more and more of the world is developed for people, there are fewer and fewer places for animals to live, eat, get fresh water, and have babies. It must be so terrifying to fly for hundreds of

Santa Barbara, 2022

miles and never see a place to land for food. We can fix that. Right this very second, if you want.

To help solve the pollution crisis, simple things can also make a big difference. Pick up litter. I go every so often with my pool net I got at a local store and scoop plastic out of the local harbor. It is a little gesture—but I believe it makes a difference to the local ecosystem—and it keeps plastic from accumulating in the oceans. I also pick up huge pockets full of trash every time I go to the beaches and mountains. These moves are like trying to stop a tidal wave, but they are first steps as part of a bigger struggle. You can remove plastic little by little from your life and start to reuse whatever you already have.

Eventually, it is going to be necessary for the companies that manufacture plastic to be responsible for tracking and cleaning up pollution from their products. The idea that an industry could release toxic or dangerous, non-biodegradable materials into the world and walk away is a recent corporate privilege, and we must do better. Some cleanup efforts are working to identify the brands with the highest environmental pollution, including the biggest food conglomerates, and require that they clean up after themselves. Corporate responsibility for site cleanup and product waste can be an essential part of our future.

So, we can break the above information into three general categories:

Civil action—Advocating for change with your government and representatives, keeping multi-national corporations on defense and voting. Folks are also suing their governments to require more protection from the impacts of climate change, and some of these court cases have been found in their favor.[1]

Personal action—Planting a garden, or pollinator plants and ripping out your lawn, providing food and water for the migrating wildlife, picking up litter and plastic wherever you go, spreading the word. The United Nations also suggests changing your diet to include less meat and more local foods, using less energy in our homes and of course, planting tons of trees![2]

Take action together—We are always stronger together, and since climate change will impact the whole world, I think working together across the whole world seems like a perfect next step. Some scholars suggest using your time, interest, and maybe even money to further the cause of fighting climate change impacts, together. Talk about climate change and biodiversity, normalize climate action, and consider your ecosystem— with your community.

There are a ton of great works by brilliant colleagues in a variety of scientific disciplines that include detailed, specific steps that you can take to work on slowing climate change. Their resources are listed below and throughout the citations in this book. This is all a part of your climate plan—and your plan to help the world can improve your local ecosystem, ameliorate any climate-anxiety, and lessen the climate crisis.

Above all, remember one thing: You can make a difference.

Put your clear eyes and open heart into this future, and we can create together and take action.

Taking action on climate change, in summary:

- You can make a difference in the overlapping biodiversity, climate change, and pollutions crises TODAY
- Collective action, personal action and civil action are all frameworks for taking action on climate change
- Fabulous resources already exist to help you plan for how to take action on climate change
- You are not alone. Reach out and take action together

A FEW RESOURCES FOR FURTHER SUPPORT

All We Can Save. https://www.allwecansave.earth.

Why We're Committing to 30x30. https://www.nature.org/en-us/what-we-do/our-priorities/protect-water-and-land/land-and-water-stories/committing-to-30x30/.

The Simple Way to Make Companies Responsible for Their Waste? Make Them Pay for It. https://www.fastcompany.com/90674483/the-simple-way-to-make-companies-responsible-for-their-waste-make-them-pay-for-it.

Actions for a Healthy Planet. https://www.un.org/en/actnow/ten-actions.

What Are the Solutions to Climate Change? https://www.greenpeace.org.uk/challenges/climate-change/solutions-climate-change/.

REFERENCES

1 U.S. Climate Change Litigation. https://climatecasechart.com/us-climate-change-litigation/.

2 10 Ways You Can Help Fight the Climate Crisis. https://www.unep.org/news-and-stories/story/10-ways-you-can-help-fight-climate-crisis.

10

WILD AND FREE

This chapter was started on the second level of a parking garage in Boston, Massachusetts, 9:45 am, November 21, Sunny, 7 degrees Celsius (45 degrees Fahrenheit), and little wind. Smells like smog and garbage.

Our lives now are a story of redemption, not restoration. It is a story of the loneliness of biodiversity loss.

There is no such thing as 'Nature.'

The famous American author, Henry David Thoreau, wrote *Walden* about living 'deliberately' in the wild woods of Massachusetts.[1] Through this elaborate tale of the strife of wild living, he neglected to mention that the time he spent in his small home in 'the wild' was only 2 miles from the nearest town—in his best friend's backyard. He often walked over to his friends' house for dinner while on his isolated sojourn.

When I found this out—standing in front of the National Landmark the house has become— I was upset. I was raised on literature from this man—congratulating and singing his praises. He was an adventurer, an explorer, and a true 'wild' man! But he was none of these things. He was a European settler, a colonialist, who lived a mile from his wealthy friend, and imagined himself something of a poet. During his time in those woods, he stole stories, time, and land from thousands who had lived with that land since time immemorial, whose experiences have never been shared as required reading for k-12 student education on multiple continents.

This perspective on the reality of a story of the 'wilderness' is important because Henry David Thoreau and many others of his time made up a story

DOI: 10.4324/9781003376323-10

Antarctica, 2013

about a country called America—a story of the colonists taming the stark and barren landscape. He described the ecosystems around him as something foreign and separate from himself and his people as something far from humans, something dangerous and to be feared and conquered. He neglected to acknowledge the Wampanoag Tribe of Gay Head (Aquinnah) and the Mashpee Wampanoag Tribes who had been living on that same land since time immemorial. Who lived there until they were killed, and the land was taken from them. The people who had called the land that Henry David Thoreau now called a 'wilderness,' a home.

This concept of the 'wilderness' has been used for hundreds of years as a justification for the subjugation of indigenous peoples globally.[2] What Henry David Thoreau and his contemporaries represented was a new take on this old story—that in this new 'American' there was a measure of strength and stamina, a fortitude of spirit,[3] required for one to leave the European-style settlements of the East Coast. We have now incorporated many of these ideas of suffering-as-merit into the dogma of progress to the detriment and destruction of our Earth. Climate change may be the ultimate symptom of this misunderstanding and calls us to change how we see the world.

These principles persisted in the Manifest Destiny ideals of early America, as colonial settlements moved into what is now the western US and Canada.[4] Today, Jeep ads that picture cars driving through roadless forests, billboards for 'adventure clothing' (covered in Teflon), and even outdoor magazines describing how to escape the city all reinforce this separation from the world around us. The thought that we should all desire to tame and conquer the 'wilderness' for ourselves. The idea that we are apart from, or better than, the world around us.[5] That humans are the most advanced, the most deserving, the most cultured, and the most intelligent species on the planet. It's a dangerous myth that has led to terrible damage.

When religions and cultures were changing in the Europe, Africa, and the Middle East, it was critical to disconnect polytheistic cultures from their gods, to move toward a monotheistic perspective.[6] This narrative continued through the development of 'naturalist' pursuits where the privileged few were able to travel from their home countries and make observations about other lands. This included Charles Darwin, whose 'Origin of the Species' has taught countless generations that the underlying structure of the world around us is competition with all things.[7] In this way it mirrors the capitalist system—that one alone must succeed above all others—and there are only few resources from which we all must draw. Quickly. Relentlessly. And with disregard for the Others. At the risk of being sacrilegious to my field, I am joining the few voices over the generations since who ask—what if Darwin was not entirely correct? What if mutualism, symbiosis—working together—is the underlying structure of the world, and we have made a grave mistake?

I remember when I learned about the nine planets of our solar system, Mercury, Venus, Earth, Mars, Jupiter, Saturn, Uranus, Neptune, and Pluto. As a pre-teen we were told that there needed to be a revision—Pluto was a planetoid, or a dwarf planet. I bring this up because it holds a visceral memory for many of my generation, and it is a perfect example of science disciplines revising and learning. The pursuit of science must be the pursuit of truth. And like all work that is in service of finding truth—when we know better, we must do better. When we know more, we must change our minds.

If the structure of competition over working together, or mutualism, isn't completely correct, we must revise our understanding, and continue to seek the truth.

The dichotomy of human versus nature, or humans as separate from the rest of the entire world (animals and plants and insects and fish and seasons and weather), is a narrative that persists and shapes many cultures to this day. It has birthed many of the challenges we now face—from mismanagement and torture of the land, deforestation, and clear-cutting to water pollution, relentless chemical dumping, and the biodiversity crisis. And of course, climate change. If one sees themselves as different, apart, owners, then cultivating a thriving environment for all creatures becomes an afterthought. However, the entire concept of separation between humans and the world around relies on outdated science. We now know that everything from the microbes in our guts to the food in our mouths to the air in our lungs is intrinsically, and forever, a part of the ecosystems we live in.

Yesterday, I saw a video of an elephant being born. All the other elephants stood in a semi-circle, watching the progress and greeting the new elephant once it had arrived. Whales are starting to spend time with whale watching boats, looking the humans above them in the eyes with understanding, an increasing occurrence now that whaling is largely banned. This morning, the ravens outside my house who go through the dumpster have sorted the trash into paper and plastic, and then look me straight in the eyes when I leave the house, they are awaiting their seed, and we know each other. We didn't realize that trees talk through vast networks of mycelia that can span miles and warn other of encroaching predators. We didn't understand that the relationship between a field mouse and the tulip that it sleeps in is mutually beneficial and rewarding. We didn't realize that elephants and whale grieve familial losses for the rest of their lives, and may even die of heartbreak, just like people.

All this is to say, that somehow, people made an error, and we found ourselves divorced from the world around us. We assumed that no other species had culture, relationships, or traditions—we even assumed they were not self-aware. And that, for me, is heartbreaking. We are surrounded by a vast, interwoven network of culturally diverse, intelligent, beautiful, and brilliant beings, who some had assumed were all mute and dumb. This is a devastating loss that has enabled much of the sprawl and disconnect in which we now find ourselves enmeshed.

We humans cherish the small creatures in our households—dogs, cats, fish, or hamsters—and this compassion and love give us a chance to regain an understanding of the larger world—of wild birds and their feelings, the culture of opossums and hedgehogs, and the pop and mourning songs of blue whales. If you listen very closely, on a clear morning, you can hear the call from this disconnected world—species using their voices and behavior to reach out to humans and ask for help, ask for peace, for some space to live their lives as well. It is the responsibility of our generations to relearn how to listen. This is also climate action.

So, as I write this, in my small home with the wind blowing through the windows and my rescue pup licking her paws on the floor beside me, I listen. I hear the noisy cars drive by the music someone has decided to share with the neighborhood, the dump truck, but I also hear the birds talking to each other. Telling stories of the wind that came from Alaska and brings cool crispness, of the warm sand and the drying of the hills, they talk of their families and what they had for breakfast. These tiny creatures that fly thousands of miles, often well above the clouds, looking for the place they grew up, driven by unseen forces of heritage and magnetic pulls, surveying the endless concrete beneath. They deserve my time, my attention, and my action.

To envision and create a new world that flourishes, we must understand the reality of what is happening around us—the creatures and their diverse cultures. As programmers work to map whale languages using AI so that we may speak with them, as birders refine what type of food to share with their local, migrating songbirds, and as we teach dogs to use buttons to speak human languages with their people, we are seeking connection and understanding. AI and other novel technologies are fallible and require energy and water we don't have to spare, but this technology is emblematic of the fact that Western societies are seeing the world differently, as an interconnected home. We have the opportunity, maybe for the first time, to use our technology and our compassion to better understand the other species on this planet. It is the contact we have **all** been looking for, and in our most disconnected time, it may save us.

What if we dreamed a world together? What if whales—the guardians and gardeners of the ocean—helped us chart a course for sustainable fisheries? What if our concrete pock-marked landscape included more vertical gardens for migrating birds to nest in? What if the water that ran through our streets after a storm was cleaned then released, so diverse fish could live in our bays and feed our children? What if we believed that this planet was not a right, but a privilege, and let it thrive. It sounds like science fiction, but weren't all of our breakthroughs once the dreams of scientists, poets, and explorers?

When we consider and envision the future, I challenge us not to hold tight to the doomsday images of endless deserts, abandoned children, and sterile shopping malls. What if instead we envisioned flower-covered alleyways, clear rivers through cities, wilderness areas protected and free, and species of all kinds and niches thriving with their families. The challenge to us now, today, at this moment in history, is to dream better.

Despite all of our searching, and all of our technology, we have only found one planet in all of the galaxies that we can see that sustain life. We may or may not be alone in the universe, but right now, on this planet, I know for sure that we are not. There are diverse networks of life all around us. There is never any reason to feel alone. Despite the threat of climate change, today the birds are singing, the sealions in the harbor are talking (almost yelling) to each other, the microbes in the soil are sending messages through root networks of flower and trees in bloom.

Generations before me wanted to make money, to have success, and the ensure the safety of their loved ones. In many ways it was selfless, and we have the opportunity to continue their example and create a world that is safe for all creatures, where we don't have to worry about the health of every songbird, the clarity of every drop of water.

But to achieve this, we must view the world differently, and we have to see our permanent place within it. I want to live in a world where I can trust that we are moving in a direction that benefits all life. A world that benefits me and my family and revitalizes the dry riverbed or restored the wild oaks on the hills. What an extraordinary thing, the idea of the sanctity of all life, and a preservation of spirit of the Land. Just snow on the mountains and water we can all drink.

Today, we are not alone on this planet and there is time and hope, for the future. Today, there is time to take action. Today, there is time to make this right. I know that if I were to be alone, on a giant spaceship, travelling 67,000 miles per hour through the vacuum of space to colonize some other world, I would rather wish to be here on this diverse, colorful, brilliant, rich ship of planet Earth. We have time, but we must act now.

Every species, every culture, every family on Earth depends on us. **Go do some Good.**

When I think about the future, this is what I wish for:

- Tree-lined walking and biking spaces
- Listening to the words of all species
- Rivers in the middle of cities alive with fish and vegetation

- Local food systems and safe housing
- Wild spaces and a pollution-free ecosystem

REFERENCES

1 Walden Pond in the Walden Pond State Reservation (U.S. National Park Service). https://www.nps.gov/places/walden-pond-in-the-walden-pond-state-reservation.htm.

2 Wulf, A. *The Invention of Nature: Alexander von Humboldt's New World* (Vintage Books, New York, 2015).

3 Westward Expansion (1801–1861) | The American Experience in the Classroom. https://americanexperience.si.edu/historical-eras/expansion/.

4 America's Manifest Destiny | The American Experience in the Classroom. https://americanexperience.si.edu/historical-eras/expansion/pair-westward-apotheosis/.

5 Reader, T. M. P. The Myth of a Wilderness without Humans. *The MIT Press Reader* (2019). https://thereader.mitpress.mit.edu/the-myth-of-a-wilderness-without-humans/.

6 Kirsch, J. *God against the gods: The History of the War between Monotheism and Polytheism* (Viking Compass, New York, 2004).

7 Darwin, C. *On the Origin of Species by Means of Natural Selection, or Preservation of Favoured Races in the Struggle for Life* (John Murray, London, 1859).

INDEX

For Product Safety Concerns and Information please contact our EU
representative GPSR@taylorandfrancis.com
Taylor & Francis Verlag GmbH, Kaufingerstraße 24, 80331 München, Germany

www.ingramcontent.com/pod-product-compliance
Lightning Source LLC
Chambersburg PA
CBHW071750270326
41928CB00013B/2865